A Dictionary
of Botany

A Dictionary of Botany

R. John Little
Claremont Graduate School

C. Eugene Jones
California State University, Fullerton

Illustrations by Raymond B. Smith

VNR VAN NOSTRAND REINHOLD COMPANY
NEW YORK CINCINNATI TORONTO LONDON MELBOURNE

Copyright © 1980 by Van Nostrand Reinhold Company Inc.

Library of Congress Catalog Card Number: 79-14968
ISBN: 0-442-24169-0
ISBN: 0-442-26019-9 pbk.

Manufactured in the United States of America

Published by Van Nostrand Reinhold Company Inc.
135 West 50th Street, New York, N.Y. 10020

Van Nostrand Reinhold Publishing
1410 Birchmount Road
Scarborough, Ontario M1P 2E7, Canada

Van Nostrand Reinhold
480 Latrobe Street
Melbourne, Victoria 3000, Australia

Van Nostrand Reinhold Company Limited
Molly Millars Lane
Wokingham, Berkshire, England

15 14 13 12 11 10 9 8 7 6 5 4 3 2 1

Library of Congress Cataloging in Publication Data

Little, R. John
    A dictionary of botany.

    1. Botany—Dictionaries.   I. Jones, C. Eugene, joint
author.   II. Title.
QK9.L735   581'.03   79-14968
ISBN 0-442-24169-0
ISBN 0-442-26019-9 pbk.

To our wives, Cynthia Little and Teresa Jones.

Preface

This dictionary contains about 5500 definitions of terms from all fields of botany and is written for students at all levels of expertise. It differs from other botanical dictionaries in several respects: a majority of the definitions included have been revised or have been rewritten from original sources; a number of new terms are included which appear for the first time in a dictionary; archaeic and seldom used terms have mostly been omitted; terms which differ slightly in ending but which have the same basic meaning (e.g., calcicole, calcicolous, calciphilous), are defined together; most taxonomic names and common names have been omitted as these words are extensively compiled in other sources; and structural formulas of chemical compounds have not been included.

In an effort to decide which terms to include and which to exclude, a survey was made of approximately one hundred sources of botanical literature including textbooks, specialized works, journals, glossaries and indices. In writing a new dictionary of botany, this approach was taken to determine how often a given term occurs, to identify new terms which ought to be included, and to compare definitions of terms as used by different authors, since definitions of a given term can vary considerably.

Authors also vary in their choice of essentially synonymous terms, some preferring one version, whereas others prefer another. We have attempted to indicate as many synonymous and opposite terms as possible. Other terms which are related in some manner to a given word are indicated by *compare* or *contrast*. Thus, a great many terms can be cross referenced by one or more means. This can provide the reader with an insight into similarities and differences between closely related terms.

The illustrations were prepared especially for this volume; many were drawn from fresh material or slides. Illustrations that depict more than one term are cross referenced in the text by a letter and number. For example, A-18 denotes the eighteenth figure in the A's, the archegonium. Thus A-18, which follows the definition of venter, indicates where this term is illustrated.

We would like to thank the many individuals who have provided criticisms, suggestions and support during the preparation of this book. We would particularily like to thank Cynthia Little for her extensive compilations of reference material and for help in proofreading. We appreciate the help of Margaret Maas, Karen Bell and Beverly Casey in typing the manuscript. A special thanks is also due to our illustrator, Ray Smith, for his many fine drawings.

R. JOHN LITTLE   C. EUGENE JONES
Claremont, Calif.   Fullerton, Calif.

A Dictionary
of Botany

A

a-. A prefix meaning without, or not.

A. **1.** Mass number of an atom. **2.** A haploid set of autosomes.

Å. See *angstrom*.

ab-. A prefix meaning from.

abaxial. Situated facing away from the axis of the plant, as the undersurface of a leaf. *Syn.* dorsal. *Opp.* ventral.

aberrant. Differing more or less widely from the accepted type; abnormal.

abiogenesis. The doctrine that living things originated from nonliving matter; spontaneous generation. Contrast biogenesis.

abiotic. Pertaining to nonliving things.

abiotic selection. Selection resulting from the interaction of an organism and its physical environment.

abjection. The shedding or throwing of spores from a sporophyte.

abjoint. To delimit by a septum or joint; to separate at a joint.

abney level. A small, handheld instrument used for determining the degree of slope.

abortive. Imperfectly developed, defective, barren.

abrupt. Terminating suddenly as though cut off.

abscisic acid. A plant growth regulator which promotes leaf abcission, prevents flowering of certain species, promotes flowering of others, and inhibits germination of several with nondormant seeds.

absciss, abscission. To fall off, as with leaves, flowers, fruits, or other plant parts.

absciss layer, abscission layer, abscission zone. One or more layers of cells that undergo transverse divisions in a zone extending across the petiole resulting in the shedding of the plant part. Layer occurs near the point of attachment of organ to plant. (*See* fig. A-1).

1

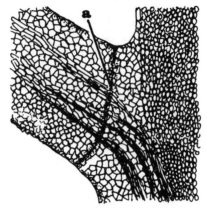

Fig. A-1. Abscission layer.

abscissa. The x-coordinate of a coordinate system; represents the distance of a point from the y-axis (ordinate) as measured parallel to the x-axis.

absolute constancy. Referring to the number of releves (sample plots) in which a given species occurs.

absolute requirement. Said of plants which will remain vegetative indefinitely and will not flower unless exposed to a required interval of low temperature.

absolute zero. Approximately −273.16°C. It is the hypothetical point on the Kelvin scale of temperature at which random molecular motion ceases. It represents the complete absence of heat.

absorption. The process of passing into, as in the absorption of water and dissolved minerals into a cell or root.

absorption spectrum. The spectrum of light waves absorbed by a particular substance, e.g., a pigment. It is measured by a spectrophotometer.

abstriction. The separation and discharge of a part as in the formation of spores or conidia in various fungi.

abundance. An estimation of the number of individuals of a species in a region.

abyssal. Pertaining to great depths. Usually used in reference to ocean depths 1829 meters (6000 feet) or more below the surface.

acantha. A spine, prickle, or thorn.

acantho-.   A prefix meaning thorny or spiny.

acanthocarpous.   Having fruit covered with spines, prickles, or thorns.

acanthocephalous.   Having a hooked beak.

acarodomatia hairs.   Hairs found in cavities; hairs found on leaves in the axils of the veins of the first and sometimes second or higher orders of ramification.

acarophyta.   Plants which harbor mites.

acarpelous.   Lacking carpels.

acarpic, acarpous.   Lacking or not producing fruit.

acaulescent.   Stemless, or apparently so; sometimes the stem is underground or protrudes only slightly; a descriptive rather than a morphological term. Contrast caulescent.

accessory.   Additional to the usual numbers of plant organs, as in accessory buds, or accessory branches.

accessory cell.   *See* subsidiary cell.

accessory chromosome.   **1.** A sex-determining chromosome. **2.** B chromosome.

accessory fruits.   **1.** Parts not derived from the ovary but associated with it when it forms a fruit. **2.** A fruit in which the major portion consists of tissue other than ovary tissue; e. g. apples, pears, strawberries.

accessory organs.   The calyx and/or the corolla.

accessory pigment.   A pigment that absorbs light energy and transfers it to chlorophyll *a*. Chlorophylls *b, c, d* are sometimes considered accessory pigments.

acclimation, acclimatization.   The natural process of adapting to a climate which was harmful at first; the process of becoming acclimated.

accrescent.   Increasing in size with age; especially any increase in calyx size after pollination.

accumbent.   Lying against and face to face; lying in contact; reclining.

accumbent cotyledons.   Cotyledons with edges lying against the embryo or the radicle.

accumulation.   The process of active uptake of ions into the root system of a plant.

-aceae.   A suffix used in names of plant families.

acellular.   Lacking cells; not composed of cells.

acentric.   Lacking a center or centromere, as in a chromosome or a chromatid.

acephalous.   Lacking a head.

acerate, acerose.   Having the shape of a needle; needlelike.

acerb.   Sour, bitter, and harsh to the taste, as with unripe fruit.

acervate.   Occurring or growing in tufts, heaps, clusters or cushions.

acervulus, *pl.* **acervuli.   1.** Discoid or pillow-shaped fruiting structures in certain fungi on which conidia are produced on conidiophores. (*See* fig. A-2). **2.** A flower cluster which is only found in the chamaedoreoid group of palms; a form of a cincinnus.

acetabuliform.   Shaped like a shallow saucer; used to describe fructifications of some lichens.

acetyl coenzyme A.   A compound formed from pyruvic acid, which then enters the Krebs Cycle; it is used in the synthesis of fatty acids and many other metabolic products.

achene.   A simple, dry, one-celled, one-seeded, indehiscent fruit; seed coat is not attached to the pericarp.

achenecetum.   An aggregation of achenes, as in *Ranunculus*.

achenodium.   A double achene, as the cremocarp of the umbel family (Apiaceae).

achlamydeous.   Lacking a perianth; without calyx or corolla.

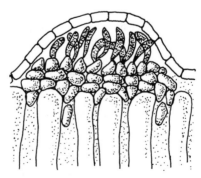

Fig. A-2. Acervulus.

achromatic apparatus, achromatic figure. In mitosis, the spindle fibers and cell centers, which do not stain readily.

acicula, *pl.* **aciculae.** A needlelike spine or bristle, as in the bristlelike continuations of the rachilla of grasses.

acicular. Needle-shaped, as an acicular leaf; having needlelike projections. as an acicular crystal.

aciculate. Marked with fine lines, usually randomly arranged.

acid. A substance having a pH below 7.0; dissociates and releases hydrogen (H^+) ions. Compare base.

acid plant. Plants which grow particularly well in acid soil.

acid soil. Soil which has a pH less than 7.0.

acorn. The fruit of the oak tree (*Quercus*), composed of a nut and its cup or cupule.

acquired characteristics. According to Lamarck's Theory of Use and Disuse, changes in environment cause modifications of structure or function in organs and resulting changes are inheritable. There is no evidence, however, to support this theory.

acra-. A prefix meaning at the apex.

acrandry. Where antheridia are found at the apex of the stem, as in bryophytes.

acranthous. Having an inflorescence borne at the tip of the main axis.

acrasin. A chemotactically active substance produced by amoebae of cellular slime molds which causes their aggregation.

acrid. Sharp, irritating or biting to the taste.

acritarchs. Name applied to microfossils whose affiliations are unknown; many acritarchs are probably fossil algae.

acro-. A prefix meaning topmost; the tip.

acrocarpic, acrocarpous. A growth form in which the gametophyte is erect and the sporophyte terminates the main axis, as in certain mosses. Compare pleurocarpous.

acrocaulous. At the tip of a stem, as a flower or fruit.

acrocentric. A chromosome or chromatid having a nearly terminal centromere.

Fig. A-3. Acrodromous.

acrocidal capsule.   A capsule that dehisces through terminal slits.

acrodrome, acrodromous.   A leaf with the main veins parallel and united at the apex, as in plantain (*Plantago*). (*See* fig. A-3).

acrogamous.   Plants in which the egg apparatus is produced at the summit of the embryo sac, as in most angiosperms.

acrogen.   A plant of the highest class of cryptogams, including the ferns, mosses, and liverworts, which have the growing point at the summit or apex.

acrogenous.   Growing from the apex (as the stems of ferns and mosses); borne at the tips of hyphae.

acrogynous.   Developing at the tip of the main shoot, with reference to the archegonia of certain scale mosses.

acrolaminar.   Ethereal oil glands located near the base of leaves.

acropetal, acropetal development.   Ascending; developing or blooming in succession from a basal position toward the apex. Said of organs (e.g. leaves, flowers), tissues or cells which develop successively on an axis so that the youngest arise at the apex. *Opp.* basipetal.

acropetalous.   An indeterminant inflorescence.

acropetiolar.   Referring to ethereal oil glands found near the apex of a petiole.

acrophytes.   Plants of alpine areas.

acroramous. Leaves positioned terminally, near the apex of a branch.

acroscopic. Facing towards the apex.

acrospire. The first shoot or sprout of a germinating seed.

actinomorphic. Flowers which are radially symmetrical; i. e. capable of being bisected into identical halves along more than one axis, forming mirror images. *Syn.* regular, radial symmetry. *Opp.* zygomorphic.

actinostele. A protostele with vascular tissue arranged in radiating star-shaped arms interspersed with parenchyma.

action spectrum. The range of light waves that elicit a particular plant response.

active absorption. Absorption of water and other materials involving an expenditure of cellular energy. Compare imbibition and diffusion.

active buds. Those which are actively growing, in contrast to dormant buds.

active site. The site on the surface of an enzyme molecule where substrate molecules are transformed into reaction products.

active transport. An energy-requiring process by which a cell moves a substance (e.g. nutrients) across the plasma membrane, often against a concentration gradient, replacing what is used in food manufacturing or other respiratory activities. (*See* fig. A-4).

active trap. In carnivorous plants, a trap in which at least a portion moves during the trapping process. Contrast passive trap.

active water absorption. The absorption of water by roots, brought about by osmotic pressure resulting from a high solute concentration in the xylem sap and permeability of the living root cells.

aculeate. Prickly, thorny, armed with spines.

aculeus. A prickle or spine, such as on a rose; a sharp epidermal emergence.

acumen. A tapering point.

acuminate. Having a long, slender, sharp point with a terminal angle less than 45°; margins straight to convex.

acute. Sharp-pointed, with a terminal angle between 45° and 90°; margins straight to convex.

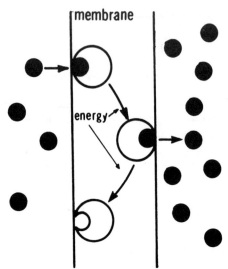

Fig. A-4. Active transport.

acyclic. Arranged in spirals, not in whorls.

ad-. A prefix meaning to, or toward.

adaptation. The process by which an organism becomes modified by a change in structure, form or function which enables it to better survive under given environmental conditions; also the result of this process.

adaptive peak. The highest degree of adaptation exhibited by an organism to a particular set of environmental parameters.

adaptive radiation. An important evolutionary process, wherein due to disruptive selection processes, a population acquires a new complex of adaptive characters which enable it to more efficiently compete for existing habitats or newly created ones, as on an island.

adaptive value. *See* fitness.

adaptive zone. The habitat or niche of an organism.

adaxial, adaxial surface. The side toward the axis; the surface of a leaf that faces the stem during development, e. g. the upper side of the leaf. *Syn.* ventral. *Opp.* dorsal.

adelphous. Having stamens united by their filaments.

adenine.   A purine base found in DNA, RNA and nucleotides such as ATP and ADP.

adenosine diphosphate (ADP).   A nucleoside diphosphate of adenosine which is involved in the transfer of energy during respiration.

adenosine monophosphate (AMP).   A compound formed by the hydrolysis of adenosine diphosphate (ADP).

adenosine triphosphatase, ATPase.   The enzyme that hydrolyzes ATP to form ADP and inorganic phosphate and catalyzes the reverse reaction of ATP formation.

adenosine triphosphate (ATP).   The nucleoside triphosphate of adenosine closely related to ADP. A high energy intermediate in energy transfer reactions, and a major source of useable chemical energy in metabolism. Most frequently found in mitochondria.

adherent.   A condition where two dissimilar organs or parts touch each other connivently but are not grown or fused together. Compare adnate.

adhesion.   **1.** The growing together or union of organs or parts which are normally separate. **2.** The attraction of unlike molecules to each other, e. g. water to cellulose.

adiabatic.   Refers to a thermodynamic process during which no heat is added to or taken from the body or system involved. No heat exchange occurs between the system and the environment.

adiabatic cooling.   The cooling of warm, low elevational air as it rises and expands; a factor contributing to cooler alpine temperatures.

adnate, adnation.   The union of unlike parts; organically united or fused with another dissimilar part, e. g. an ovary to a calyx tube, or stamens to petals. Compare adherent.

ADP.   *See* adenosine diphosphate.

adsorption.   Adhesion of the molecules of a gas, liquid or dissolved substance to a solid surface or interface; it consists of an interfacial concentration of molecules, and occurs in colloidal and noncolloidal systems.

advanced.   A term applied to an organism or part of an organism which implies considerable difference or modification from the ancestral condition.

adventitious.   Plant organs produced in an unusual or irregular position, or at an unusual time of development, as in adventitious buds, roots or shoots. (*See* fig. A-5).

adventitious embryo, embryony.   An embryo formed without fertilization, which de-

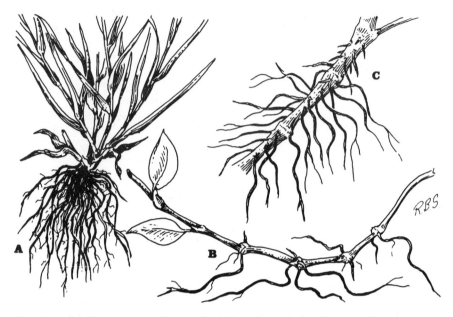

Fig. A-5. Adventitious roots: A. originate at base, B. originate at nodes, C. originate between nodes.

velops directly from the ovular tissue of the parental sporophyte, usually from the integument of the nucellus.

adventive.   Introduced but not well established; recently or incompletely naturalized, or temporarily established.

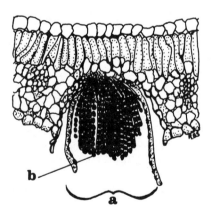

Fig. A-6. Aecium in leaf cross section: a. aecium, b. aeciospores.

aecial initial.   A coil of hyphae produced by the haploid mycelium resulting from the germination of a basidiospore of some rust fungi.

aeciospore.   A spore formed in an aecium; a binucleate spore of a rust fungus. (*See* fig. A-6).

aecium, *pl.* **aecia.**   A cup-shaped structure consisting of dikaryotic hyphal cells produced by some rust fungi on certain host plants. (*See* fig. A-6).

aerating roots.   Roots which arise from the soil surface as from mud. Such roots have loose corky tissue (aerenchyma) with conspicuous intercellular spaces thought to function in aeration.

aerating tissue.   *See* aerenchyma.

aeration.   The entrance of air; the mixing with air, or being supplied with air.

aerenchyma.   Parenchyma tissue characterized by particularly large intercellular spaces; respiratory tissue formed by the phellogen. *Syn.* aerating tissue.

aerial.   Living above the surface of the ground or water.

aerial plants.   Plants not rooted in soil, such as epiphytic plants; e.g. bromeliads (*Tillandsia*), and some tropical orchids.

aerial roots or stems.   Roots or stems growing from adventitious buds into the air; roots which develop partly or entirely above the ground, as the roots of orchids.

aerobe.   An organism which requires free oxygen for maintenance of life.

aerobic.   Capable of living only in the presence of free oxygen.

aerobic respiration.   Respiration occurring in the presence of oxygen.

aerobiology.   The study of small organisms and their products present in the atmosphere; e.g. spores and pollen.

aerocaulous.   Referring to plants with aerial stems.

aerola, *pl.* **aerolae.**   Wall markings in certain diatoms, consisting of thin areas bounded by ridges of siliceous material and having an aggregation of many fine pores.

aerophyllous.   Referring to plants with aerial leaves.

aerophyte.   A plant that grows attached to an aerial portion of another plant, obtaining water from rain or dew, as an orchid; an epiphyte.

aestival.   With flowers appearing in summer.

aestivation, estivation.   **1.** The arrangement of the perianth or its parts in the bud. Compare vernation. **2.** Passing through a hot dry season in an inactive state.

aethalium, *pl.* **aethalia.**   A sessile, rounded, generally pillow-shaped fructification formed by a massing of all or most of the plasmodium in some of the slime molds (Myxomycetes).

affinis.   Closely related to another.

affinity.   Similarity between entities in regard to morphological traits.

afforestation.   To convert into a forest or woodland, usually where trees have not previously grown.

after-ripening.   The metabolic changes that must occur in some dormant seeds before germination can occur.

agamic, agamous.   Asexual; reproduction without union of sex cells.

agamic complex.   An asexually (apomictically) reproducing population of plants.

agamogenesis.   Asexual reproduction of any kind, e. g. by buds, gemmae, etc.; without formation of functional gametes; parthenogenesis.

agamospecies.   An apomictic population believed derived from a common ancestor based on morphological, cytological and other data.

agamospermy.   The production of seeds by asexual means, i. e. without fertilization.

agar.   A non-nitrogenous, gelatinous mixture of polysaccharides obtained from certain red algae. Used extensively as a solidifying agent in laboratory culture media for bacteria, fungi and tissue cultures.

agenesis.   Failure to develop.

agents.   Organic or inorganic means which effect seed dispersal such as wind, water and animals; also vectors of pollination such as wind, insects, etc.

aggregate.   **1.** Clustered together to form a dense mass or head, usually applied to an inflorescence. **2.** Gravel, crushed rock, or other coarse, inert material that can be mixed with soil to make it more porous.

aggregate cup fruit.   A fruit derived from an apocarpous (free carpels), perigynous flower; fruit composed of fruitlets, each with its own pericarp.

Fig. A-7. Aggregate fruit: a. seed, b. follicle.

aggregate free fruit.   A fruit derived from an apocarpous (free carpels), hypogynous flower; fruit composed of fruitlets, each with its own pericarp.

aggregate fruit.   A fruit developed from a single flower by fusion of many separate carpels (pistils), all of which ripen together into one mass, as in blackberry, raspberry, and magnolia. (*See* fig. A-7).

aggregate ray.   In secondary vascular tissues, a group of small rays arranged so as to appear to be one large ray.

aggregation.   1. The condensation of cell contents under some stimulus; the movement of protoplasm in tenacle or tendril cells of sensitive plants which causes the tendril to bend toward the point stimulated. **2.** A group or mass of individuals. **3.** The movement toward a single point of amoebae in some cellular slime molds (Acrasiomycetes) prior to pseudo-plasmodium formation.

agmatoploidy.   Used in reference to differences in the number of independently assorting pairs of chromosomal fragments which occur in a group of organisms. Characteristic of organisms lacking a localized centromere; i. e. they are polycentric.

agricultural ecosystem.   An ecosystem based on the cultivation of plants and animals by man.

agrobiology.   The study of plant nutrition, crop production, and growth in relation to soil control.

agrology.   The branch of agricultural science which deals with the study of soils.

agronomy.   The study and practice of field crop production and soil management.

agrostology.   The branch of systematic botany dealing with the study of grasses.

agynic. Flowers which lack pistils.

A-horizon. The upper layer of soil composed of various types of decomposing organic materials, such as leaf litter, raw humus, and decomposed humus resting on residual soil. *Syn.* A-layer.

aianthous. Blooming continuously; having everlasting flowers.

aigrette. Any feathery crown or tuft attached to a seed, as an adaptation for wind dissemination, as the pappus of a dandelion fruit.

air bladder. Organs filled with air which give plants buoyancy in water, as those of certain brown algae.

air chamber. An anatomical cavity which contains air. (*See* fig. M-6).

air fern. The dead skeleton of a marine animal of the genus *Bugula*. Because of its fernlike growth when dried, it is sold as a novelty as a "living fern", after being dyed green.

air plants. Epiphytes; plants growing in the air above the ground.

air roots. *See* pneumatophores.

air sacs. Cavities containing air; as those in the appendages (wings) of the pollen grains of most pines (*Pinus*).

air spaces. Intercellular spaces.

akaryote. A phase in the life cycle of certain primitive fungi during which the nucleoplasm loses its affinity for stains.

akinete. A vegetative cell of certain green algae whose original walls thicken and become transformed into a nonmotile or resting spore.

ala, *pl.* **alae.** One of the side petals of a papilionaceous flower (as in the pea and its relatives); a wing or winglike process, as the bladelike expansion along the axial margin of certain carnivorous plant leaves.

alar cells, alar region. Cells at the basal angles of a moss leaf or scale, which differ from the rest of the leaf in size, shape or color.

alar region. Cells in the basal corner of a moss leaf.

alate. Winged, as in certain stems, or seeds; having winglike extensions.

A-layer.   *See* A-horizon.

albedo.   In citrus fruits, the white tissue of the rind.

albino.   A plant having colorless chromatophores due to the lack of chlorophyll in chloroplasts.

albumen, albuminous.   Starchy or other nutritive material within the embryo; an older term largely replaced by the term "endosperm".

albuminous cells.   Parenchyma cells in gymnosperm phloem and ray that appear closely related with the sieve elements, both morphologically and physiologically, but are not derived from the same initials. *Syn.* strasburger cells.

albuminous seed.   A seed that contains endosperm at the time of germination, supplying nourishment to the growing seedling.

alcohol.   An organic, liquid compound, colorless and inflammable; e. g. ethyl alcohol.

alcoholic fermentation.   An anaerobic process whereby alcohol is produced from the action of yeast on a sugar substrate.

-ales.   A suffix denoting taxonomic orders of plants.

aleurone.   Granules of protein and enzymes present in the ripe seed of numerous plants; usually pertaining to the external part of endosperm of wheat or other cereals.

aleurone layer.   The outermost layer of endosperm in grains and many other taxa. (*See* figs. A-8, R-1).

alga, *pl.* **algae.**   Photosynthetic eukaryotic organisms that generally reproduce by unicellular sex organs; the blue-green algae are photosynthetic prokaryotes. (Note: although in common usage, the term "blue-green algae" should probably be avoided because of the implication of relationship with the true algae). *See* blue-greens.

algal layer.   A layer of green algal cells lying inside the thallus of a heteromerous lichen, also called the gonidial layer or algal zone.

algin.   A commercially valuable gelatinous substance obtained from the cell walls of brown algae, and often used as a thickening agent in the production of ice cream.

aliform paratracheal parenchyma.   Vasicentric groups of axial parenchyma having tangential winglike extensions as seen in cross-section. Found in secondary xylem.

aliphatic.   Organic compounds which are composed of linear chains of carbon atoms. Compare aromatic and heterocyclic.

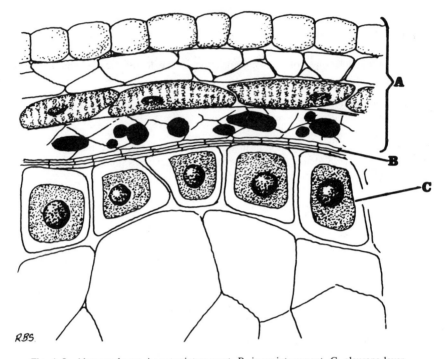

Fig. A-8. Aleurone layer: A. outer integument, B. inner integument, C. aleurone layer.

aliquot.   A portion; a sample which evenly divides the whole, leaving no remainder. Loosely used as any fraction or part of a whole.

alkaline.   Above pH 7.0; said of soil, when describing the distribution of certain types of plants which are alkaline tolerant. A base.

alkaloids.   Organic compounds produced by plants which have alkaline properties and contain carbon, hydrogen, nitrogen; these form the bases of the active portion of many drugs and plant poisons.

allagostemonous.   Having stamens attached to petal and receptacle alternately.

allantoid.   Sausage-shaped.

allautogamia.   The ability of plants to be pollinated by two methods, one usual, the other facultative.

allele, allelomorph.   One of the two or more alternative forms of a gene that occurs at the same locus on homologous chromosomes; because alternate forms exist, an organism may be homozygous for some alleles and heterozygous for others, and would then be subject to Mendelian inheritance.

allelopathy. Biochemical inhibition between higher plants, and between higher plants and soil microorganisms; caused by the release of various metabolic substances such as terpenes, camphor, and cineole.

allo-. A prefix meaning other or different.

allochthonous. A native species that did not originate in an area but immigrated into it and now reproduces naturally.

allogamous. Requiring two individuals to accomplish sexual reproduction; habitually cross-fertilized although capable of self-fertilization.

allogamy. Cross-fertilization; opposite of autogamy.

allometry. The relationship between or among differing growth rates in different parts of the same organisms.

allopatric, allopatry. Species of populations orginating in or occurring in different geographical regions, without overlapping distributions. Compare sympatric.

allopatric speciation. Speciation in geographical isolation.

allophene. A phenotype due to other than the mutant genetic constitution of the cells of the tissue involved. Compare autophene.

allophilic. Refers to a flower available for pollination by any animal visitor.

allophycocyanin. A blue biliprotein pigment of the blue-greens (Cyanophyta).

alloploid, allopolyploid. A polyploid having more than two sets of chromosomes which are genetically different; e.g. allotetraploid, allohexaploid; this results from hybridization between different species. *Syn.* amphiploid.

allosomes. Sex chromosomes.

allosteric enzymes. Enzymes which are capable of reacting with certain compounds so that their enzymatic action is either promoted or inhibited.

allosyndesis. In polyploids and their hybrids, the pairing of chromosomes derived from different (opposite) parents.

allotropic. Flowers which are poorly adapted to insect visitors, as a means of affecting pollination.

allozymes.   Genetic variants of various enzymes.

alluvial soils.   A type of azonal soil which is highly variable and is classified by texture from fine clay/silt soils through gravel and boulder deposits.

alluvium.   Soil, usually rich in minerals, deposited by water, as in a flood plain.

alpine.   1. Refers to plants growing above the timber line. **2.** An ecological life zone.

alternate.   Any arrangement of leaves or other plant parts which are not opposite or whorled but are placed singly on the axis or stem at different heights.

alternate host.   One of the two hosts of a heteroecious rust fungus; often refers to the host of less economic importance.

alternate pitting.   A type of pitting in tracheary elements, in which the pits are arranged in diagonal rows. (*See* fig. P-9).

alternation of generations.   The alternation of gametophytic and sporophytic generations in the life cycle. The sporophyte develops from the zygote and produces spores. The gametophyte develops from the spore and produces gametes. Cells of the gametophyte have 1n number of chromosomes, whereas the sporophyte has 2n. (*See* fig. A-9).

alveolate.   Honeycombed; having deep angular cavities separated by thin partitions.

alveolus, *pl.* **alveoli.**   A pit or depression with angular sides, as the cavity of a honeycomb.

amber.   Fossilized resin from prehistoric conifers.

ambient.   Relating to the environment at a particular time, as in ambient temperature.

ambiguous.   Said of plants when their taxonomic position of classification is doubtful.

ameiotic.   Division of nuclear material by cleavage or splitting without meiosis or visible chromosome formation.

ament.   A spike of apetalous flowers, pistillate or staminate, as in oaks or willows; a catkin.

amentaceous.   Bearing catkins.

amino acids.   The primary components of proteins, each containing an amino group and linked together by peptide bonds; large numbers of amino acids connected together in this manner form proteins.

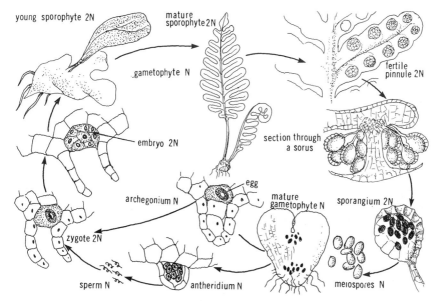

Fig. A-9. Alternation of generations.

amitosis, amitotic.   Division of nuclear material by cleavage or splitting without mitosis or visible chromosome formation; the result being that the chromosomes may not be equally distributed or a cell may contain two or more nuclei.

amixis.   Where interbreeding fails to occur between different species.

ammonification.   The production of ammonia (NH_3) and ammonium (NH_4^+) by bacterial organisms of the soil acting upon nitrogen-containing organic compounds.

ammonifying bacteria.   Bacteria which decompose proteins into ammonia-forming ammonium compounds in the soil.

amoeboid.   **1.** Resembling an amoeba; protoplasm showing a creeping movement similar to that of the plasmodium in the Myxomycetes (slime molds). **2.** Without a cell wall.

amorph.   A mutant allele having no effect when compared to the wild-type allele.

amorphic, amorphous.   Without definite form, or shape.

AMP.   *See* adenosine monophosphate.

amphi-.   A prefix meaning around, on both sides or of both types.

amphibian, amphibious.   A plant adapted to live either on land or in water; the ability to live in either environment.

amphicarpous.   Producing two kinds of fruit.

amphichrome, amphichrony.   Producing flowers of two different colors on the same stem.

amphicribral vascular bundle.   A concentric vascular bundle in which the phloem surrounds the central strand of xylem.

amphiflorous.   A plant with flowers or fruits above and below ground.

amphigastrium, *pl.* **amphigastria.**   The ventral leaf of leafy liverworts (Hepaticae), which is often reduced in size and resembles stipules.

amphigenous.   Growing over or around an entire surface.

amphimixis.   True sexual reproduction; an embryo is formed from the union of male and female gametes, each with a different set of chromosomes. Compare apomixis.

amphiphloic siphonostele.   A stele having a pith and two phloem regions, one outside the cylinder of xylem and one on the inside; having phloem on both sides of the xylem.

amphiploid, amphidiploid.   A hybrid between species which contain different chromosome numbers, the new total being the sum of both parents; this hybrid can arise from two species of the same genera or between two different, but related genera, e.g. *Tritcale* ($n=\pm28$) from *Triticum aestivum* ($n=21$) crossed with *Secale cereale* ($n=7$). *Syn.* alloploid. Compare autoploid *(See* fig. A-10).

amphisarca.   A berrylike succulent fruit surrounded by a woody or crustaceous layer or rind forming a gourd, as in *Lagenaria* (Bottle gourd).

amphistomatic.   Leaves which have stomata on both surfaces.

amphithecium.   The external layer of cells surrounding the sporogenous tissue in the sporangium of a moss.

amphitrichous.   Having flagella at both poles of a bacterium.

amphitropous ovule.   An ovule which is attached near its middle, and is bent in the form of a "U", so that the ovule tip and stalk base are near each other.

amphivasal, amphivasal vascular bundle.   A bundle with the xylem surrounding a strand of phloem. *Syn.* concentric.

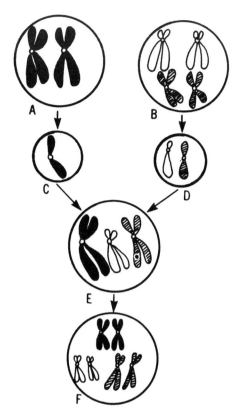

Fig. A-10. Amphidiploid formation resulting from hybridization and chromosome doubling between steps E and F: A.B. parental types, C.D. gametes, E. hybrid, F. amphidiploid.

amplectant. Clasping or winding tightly around an object for support, as with tendrils.

amplexicaul. Clasping the stem, as the base of certain leaves.

ampulla. A bladder, which captures small aquatic organisms, as in the bladderworts, (Lentibulariaceae). (*See* fig. A-11).

amylase. An enzyme capable of converting starch to sugar via hydrolysis.

amyloid. Turning blue-black after addition of an iodine solution, indicating a starchlike reaction.

amylopectin. The relatively insoluable component of starch, composed of α 1-6, 1-4 glucoside linkages.

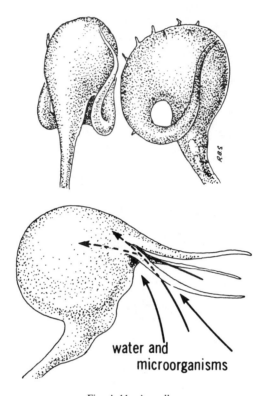

water and
microorganisms

Fig. A-11. Ampulla.

amyloplast, amyloplastid.   A colorless leucoplast in which starch is synthesized.

amylose.   The soluble component of starch, composed of α 1-4 glucoside linkages.

an-.   A prefix meaning without or not. Equivalent to the prefix a-.

anabolism.   Constructive metabolism, such as photosynthesis, assimilation, and synthesis of proteins, resulting in the building of new, or repair of old cells; also, reactions leading to the storage of energy. *Opp.* catabolism.

anacrogynous.   A condition in some liverworts (Hepaticae), in which the gametangia are borne in a lateral position, having been formed from subapical cells, the result being that the sporophyte is borne laterally.

anadromous.   Denoting the type of venation in which the first lobe or segment of a pinna arises basiscopically in compound leaves of certain ferns, e. g. *Asplenium.*

anaerobe.   An organism able to live without free oxygen (atmospheric or molecular oxygen).

anaerobe (faculatative).   An organism able to live either in the absence of free oxygen, or in the presence of it.

anaerobe (obligate).   An organism which can live only in the absence of free oxygen.

anaerobic.   Capable of living in the absence of or not requiring molecular oxygen, e. g. anerobic bacteria.

anaerobic respiration.   A type of respiration found only in bacteria; it occurs in the absence of free molecular oxygen, and results in low ATP formation; the hydrogen released is combined with the bound oxygen of inorganic compounds, e.g. iron, sulfur, or carbon dioxide.

anagenesis.   The process of evolution which leads to biological improvement, resulting in an improved type (species) which partly or completely replaces an older and less improved type; the product and functional unit of anagenesis is called a grade.

analogous structures.   Organs which appear similar because of their function, but do not have a common origin or ancestry, e.g. bird and insect wings.

analogy.   The concept that certain organs may have a similar function and appearance, yet are not related in terms of embryonic or phylogenetic origin. Contrast homology.

anandrous.   Having a perianth and pistils but lacking stamens; said of pistillate flowers.

anaphase.   The state in mitosis or meiosis (anaphase II), during which the chromatids of each chromosome separate and move from the equatorial plate to their respective poles.

anastomosis, anastomosing.   Parts joined or coming together; interconnection of elongated structures, such as veins and other cell strands forming a reticulum or netlike appearance.

anatomy.   The study of plant morphology that deals with the internal structure and form of plants.

anatropous ovule.   Said of an ovule that is inverted, so that its opening (micropyle) is near the base of the funiculus attachment.

ancestry.   Those members of past generations which are related to any given species by descent.

anchor.   A plant holdfast used for attachment to a substrate.

Fig. A-12. Anchoring organs.

anchoring organs. The ends of certain tendrils which have specialized flattened disks for attachment. (*See* fig. A-12).

ancipital. Two-edged; flattened or compressed, as in stems of certain grasses.

andragamous. An inflorescence with staminate flowers inside or above and neuter flowers outside or below.

andro-. A prefix meaning male.

androcyte. A cell which eventually develops into an antherozoid; a sperm mother cell or spermatid.

androdioecious. A species in which some plants bear staminate (male) flowers while others bear perfect (hermaphroditic) flowers.

androecium. A collective term referring to the stamens of a flower; the stamens as a unit of the flower; the male portion of a flower.

androgamete. In seed plants, the male gametes (sperm) or sex cells produced by the mitotic division of the androgametophyte within the pollen grain.

androgametophyte. 1. In seed plants, a gametophyte within the wall of the pollen grain which produces the androgametes. 2. The germinated pollen grain.

androgenesis. Development of an individual from a sperm following the disintegration of the egg prior to syngamy. Such individuals are haploid and have only paternal chromosomes, as in certain varieties of tobacco (*Nicotiana*).

androgenous. Bearing males; referring to the production of males or male gametes.

androgenous cell. *See* spermatogenous cell.

androgynecandrous. An inflorescence with staminate flowers above and with pistillate flowers below, as in the spikes of some species of sedges; e. g. *Carex.*

androgynophore. A stalk bearing both stamens and pistil above the point of perianth attachment.

androgynous. Having staminate and pistillate flowers in the same inflorescence, or cluster; an inflorescence with staminate flowers inside or above and pistillate outside or below, as in some sedges.

androhermaphroditic. An inflorescence with staminate flowers inside or above and hermaphroditic outside or below.

andromonoecious. A plant with staminate and perfect flowers, but lacking pistillate flowers.

androsporangium. In seed plants, the sporangium in which androspores are produced. *Syn.* microsporangium.

androspore. **1.** In seed plants a spore that produces the male gametophyte (androgametophyte). **2.** An asexual spore which gives rise to a dwarf male plant in certain green algae. *See* microspore.

androsporophyll. In seed plants an appendage that bears the androsporangium. *See* microsporophyll.

androstrobilus. A strobilus or cone bearing microsporangia or pollen sacs. *See* microstrobilus.

anemo-. A prefix referring to wind.

anemophilous, anemophily. Referring to flowers which are wind-pollinated. Contrast entomophilous.

anemotropism. The orientation of a plant or a part of it in response to air currents.

aneuploid, aneuploidy. An organism whose somatic chromosome number is not an exact or even multiple of the basic haploid number.

anfractuose. Wavy, sinuous, twisted; serpentine.

angio-. A prefix meaning a closed structure, capsule or locule.

angiocarpic, angiocarpy. Having an enclosed fruit.

angiosperm.   Any flowering plant; any plant of the Angiospermae; a plant bearing seeds which develop in an enclosed ovary or carpel.

Angiospermae.   One of the two large divisions of seed plants distinguished by having true flowers and ovules which are enclosed in a carpel.

angstrom, angstrom unit.   One hundred-millionth of a centimeter; one ten-thousandth of a micron; one-tenth millimicron. Symbol Å.

angular collenchyma.   Collenchyma tissue in which the cell wall thickening is deposited mainly in the angles where several cells are joined together.

angulate.   More or less angular; having angles or corners usually of a determinate number.

anion.   A negatively charged ion. Contrast cation.

aniso-.   A prefix meaning unequal.

anisocarpic, anisocarpous.   Having fewer carpels than the number of parts in the other floral whorls; having unequal carpels.

anisocotylous, anisocotyledonous.   Having unequally developed cotyledons.

anisocytic stoma.   A type where three subsidiary cells, one distinctly smaller than the other two, surround the stoma. (*See* fig. S-16).

anisogamous planogametes.   Motile gametes which are morphologically similar but differ in size.

anisogamous, anisogamy.   The fusion of two similar motile gametes differing in size, the larger gamete considered the female and the smaller the male. Compare isogamous and oogamous. (*See* fig. A-13).

anisolateral.   With unequal sides.

anisomerous.   With an unequal number of floral parts in the different whorls.

anisopetalous.   With petals of unequal size.

anisophyllous.   Having leaves of two or more sizes or shapes.

anisopterous.   Unequally winged, referring to seeds.

anisosepalous.   With sepals of unequal size.

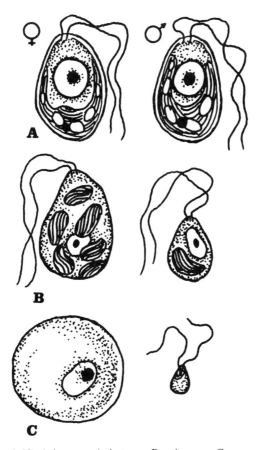

Fig. A-13. Anisogamy: A. isogamy, B. anisogamy, C. oogamy.

anisospores. Dimorphic spores with the male and female spores of different sizes.

anisostylous. With styles of unequal length.

anisotropic. With different properties along different axes; optically anisotropic; crystalline.

anlage. *See* primordium.

anneal. To heat and then cool slowly.

annotinal, annotinous. Structures which appear yearly, as leaves, flowers, etc., and which are renewed every year.

annual. A plant which completes its life cycle within one year or one growing season; i.e. it grows vegetatively, produces flowers and sets seed in one season.

annual cycle. Periodicity; the study of yearly periodicity in plants as related to climatic events, e. g. time of flowering, fruiting, or budding.

annual ring. A layer of wood (xylem) formed in the stems of woody plants each year by the vascular cambium, and appears as concentric layers when viewed in cross section. The term growth layer is preferred because more than one growth layer may be formed during one year. (*See* fig. A-14).

annual shoot. The shoot produced each year in the spring by a perennial.

annual, winter. A plant whose seeds germinate in the fall, live in a semi-dormant state during winter, produce seeds in the spring, and then die.

annular. In a ring or ring shaped; forming a ring or circle.

annular rings, annular thickening. Secondary thickenings on the walls of fiber cells, vessel elements, or tracheids.

annulus, *pl.* **annuli.** **1.** In some ferns, a specialized ring of thick-walled cells extending from the stalk over the sporangium. (*See* fig. L-5). **2.** In mushrooms, a membranous ring or remnant of the inner veil left as a ring around the stipe (stalk) after the expansion of the cap. (*See* fig. B-2). **3.** In mosses, the line of cells found between the mouth of the capsule and the operculum. **4.** Any ring or ringlike structure.

anodic. The ascending direction of a leaf-spiral.

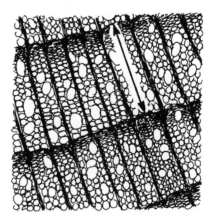

Fig. A-14. Annual ring in stem cross section.

anomalicidal capsule.   A capsule that dehisces irregularly; a rupturing capsule.

anomalous secondary growth.   Referring to unusual types of secondary growth.

anomocytic stoma.   A type of stoma in which subsidiary cells are not associated with the guard cells.

antechamber.   The space just below the guard cells of a stoma.

antepenultimate.   Third from the end.

anterior.   **1.** On the front side. **2.** In the flower, the side away from the axis and toward the subtending bract.

anthelate.   Having elongated, flower-bearing branches as in certain rushes.

anther.   The pollen-forming part of the stamen; the microsporangium. (*See* fig. F-3).

anther lobes, anther sac.   The lobes of an anther which contain pollen.

anther tube.   A tube formed by coalescent anthers, as in the sunflower family, (Asteraceae).

antheridial initial.   The cell which results from the division of an androspore cell; divides to form a generative cell and a tube cell.

antheridiophore.   A specialized structure upon which antheridia are borne, e.g. in certain liverworts. (*See* fig. H-2).

antheridium, *pl.* **antheridia.**   A male organ of reproduction producing sperms (male gametes); it consists of a single cell in fungi and algae, and is multicellular in mosses and vascular plants. (*See* fig. A-9).

antheriferous.   Bearing anthers.

antherozoid.   A small, flagellated, male sex cell (gamete), which develops in an antheridium of certain algae, mosses, and ferns; a sperm.

anthesis.   The time at which a flower comes into full bloom; when a flower is fully expanded.

anthium.   In pollination ecology, a term proposed to account for the functional concept of a pollination unit, as different from an attraction unit.

anthocarp.   A dry, indehiscent fruit composed of an ovary surrounded or united by a perianth tube or receptacle, as in the fruit of the Nyctaginaceae, (four-o-clocks).

anthocarpous.   Having a body of combined flowers and fruits which are united in a single mass, as in multiple fruits, e. g. pineapple and mulberry.

anthocyanin.   Natural pigments of blue, purple or red which are especially common in the petals of flowers; these pigments vary in color according to the acidity of the cell sap, and are composed of soluble glucosides.

anthocyanous.   Showing the presence of anthocyanin in the herbage.

anthodium.   A composite flower head; capitulum.

anthophyte.   Any member of the Anthophyta, a division of the plant kingdom which includes all angiosperms or seed-bearing plants that develop seeds in a carpel.

anthos-.   A prefix meaning flower.

anthotaxis.   The arrangement of flowers, (sporophylls), on an axis.

anthoxanthin.   Glycoside pigments chemically similar to anthocyanins, and colorless or nearly so as they occur in the plant; these pigments are sometimes yellow in certain flowers, e. g. yellow snapdragons.

anthracnose.   A common type of plant disease caused by fungi producing nonsexual spores in acervuli; characterized by limited necrotic lesions on stems, leaves, or fruits of the host, often accompanied by dieback. (*See* fig. A-15).

anthrophile, anthropophile.   A plant which usually follows cultivation or civilization; a plant of disturbed areas.

Fig. A-15. Anthracnose.

anthropochore. A plant distributed by man or involuntarily introduced and having no special adaptation for seed dispersal.

anthropomorphic. Attributing human qualities to nonhuman organisms or objects.

anthropophily. Pollination by man.

anthropophytes. Plants which exist in a habitat disturbed by man, e. g. cultivated fields, roadsides, and waste places.

-anthus. A suffix meaning flower.

antiauxins. Compounds which are chemically similar to auxins and are able to negate normal auxin activity, apparently through competitive inhibition.

antibionts. Plants which do not graft easily.

antibiosis. An association between living organisms that results in the suppression of the life of one of them.

antibiotics. Substances derived from microorganisms that interfere with the metabolism of microorganisms.

antibody. A specific protein produced by an organism in response to an antigen.

anticlinal. 1. Perpendicular to the surface; radial. 2. Said of cell divisions which occur at right angles to the surface of the meristem. Contrast periclinal. (*See* fig. P-4).

anticodon. A triplet code unit on a transfer RNA molecule which is complementary to the codon carried by messenger RNA.

antidromous, antidromy. A condition where a spiral is reversed from the usual direction.

antigen. A substance that stimulates the production of antibodies.

antimorph. A allele with an action opposite to that of the wild type.

antipetalous. Opposite the petals, as in stamens which are situated in front of petals.

antipodal. Cells or cell structures at the base or foot of the embryo sac; being furthest from the micropyle in the seed. (*See* fig. 0-3).

antipodal cells. One or more cells at the chalaza end of the seed embryo sac of angiosperms. (*See* figs. E-2, O-3).

antiserum.   A serum containing antibodies.

antitropous.   Embryos which have the radicle pointing away from the hilum.

antrorse.   Bent or directed upward or forward. *Opp.* retrorse.

aperturate.   A pollen grain with one or more apertures.

apertures.   **1.** Any preformed thinning or absence of a part of the exine of a pollen grain. **2.** An opening or hole. (*See* fig. B-6).

apetalous.   Lacking petals; said of flowers which have petals reduced to inconspicuous scales.

apex, *pl.* **apices.**   The tip; top most part or terminal end; the part of a root or shoot containing the apical meristem.

aphanoplasmodium.   A plasmodium consisting of a network of very fine transparent, undifferentiated strands in which the protoplasm is not granular.

aphlebia.   Stipular outgrowths which occur on the rachis of many fossil fern fronds.

aphotic zone or region.   That portion of a body of water which receives little or no light; e.g. generally depths below 800 meters. (*See* fig. A-16).

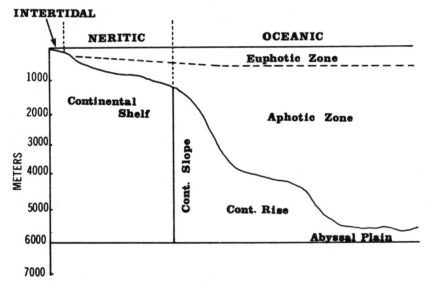

Fig. A-16. Aphotic zone.

aphyllopodic.   Plants without blade-bearing leaves at its base.

aphyllous.   Leafless, without leaves.

aphytal.   Lacking plants.

apical.   At, near, or belonging to the apex or point; of the tip of a summit.

apical cell.   The single initial meristematic cell in an apical meristem of a root or shoot which divides repeatedly to form new tissues for the plant or organ. They are found in filamentous lower vascular plants, thalli of mosses, and in root and shoot tips of higher plants. (*See* fig. S-4).

apical dominance.   The influence exerted by a terminal bud in suppressing the growth of lateral buds; this control results mainly from auxin content in the terminal bud, as well as the amount of cytokinins in the lateral buds.

apical meristem.   A group of meristematic cells at the tip of a root or shoot from which all the tissues of the mature axis are ultimately formed; these may be vegetative or reproductive cells. (*See* figs. M-2, P-15).

apicula.   A short, sharp, flexible point.

apiculate.   With a short, abrupt, or acute point; terminated by an apicula.

aplanetic.   Nonmotile; the absence of motile spores.

aplanogamete.   A nonmotile gamete.

aplanospores.   A nonmotile spore, as in the resting spores of algae and in nonciliated reproductive cells of some green algae.

apocarp, apocarpous, apocarpy.   The condition in which carpels are separate and unfused; not united; frequently applied to a gynoecium of separate pistils. Contrast syncarpous.

apochlamydeous.   With perianth parts not united or fused.

apodal.   Without a foot-stalk or podium; sessile; especially in reference to trichomes.

apogamy, apogamety.   The phenomenon shown by some higher plants in which the vegetative cells of the gametophyte give rise directly to a sporophyte without the fusion of gametes; parthenogenesis.

apogeotropic.   Growing up out of, or away from the earth; negatively geotropic.

apolar pollen.   Pollen with no distinct polarity discernible before tetrad separation.

apomictic.   Capable of producing seed without any form of fertilization or sexual union.

apomictic population.   A population resulting from asexual reproduction.

apomixis.   Reproduction, including vegetative propagation, which does not involve sexual processes; the ability of plants to produce seeds without fertilization. Compare amphimixis.

apopetalous.   Having the petals of the corolla unattached to each other; with separate and distinct petals; polypetalous.

apophysis.   **1.** An enlargement of the seta at the base of a capsule in certain mosses. **2.** The rounded, exposed thickening on the scales of certain pine cones.

apoplast.   A system consisting of all the cell walls of a plant through which water moves rather freely, including dead cells of vascular tissues.

aporogamy.   Penetration of an ovule by a pollen tube through a point other than the micropyle. *See* chalazogamy and mesogamy.

aposepalous.   With free or distinct sepals; not having the sepals of the calyx attached to each other.

apospory.   The phenomenon shown by some higher plants in which the vegetative cells of the sporophyte give rise to a gametophyte forming an embryo without fertilization. This is accomplished when a diploid embryo sac is formed directly from a somatic cell of the nucellus or chalaza.

apostemonous.   The condition wherein stamens are completely separate.

apothecium, *pl.* **apothecia.**   An open ascocarp of many lichens and certain fungi, often discoid or cup-shaped, in which the hymenium is exposed at the maturity of the asco-spores. (*See* fig. A-17).

apotracheal parenchyma.   Axial parenchyma cells in wood typically independent of the pores or vessels; these include boundary, banded and diffuse apotracheal.

appendage.   An attached subsidiary or secondary part, as a projecting or hanging organ.

appendicular (stamen).   A typical stamen type which is variously shaped or modified with a protruding connective, as in *Viola*.

appendicular theory.   A theory suggested to explain the inferior ovaries of angiosperms

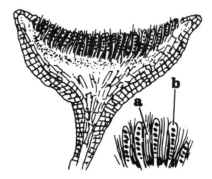

Fig. A-17. Apothecium: a. paraphyses, b. asci.

on the basis of extensive fusion of the outer floral whorls to each other and to the ovary wall; other postulated evidence comes from the course of the vascular bundles supplying the ovules.

appendix.   The sterile terminal portion of a spadix of the Araceae (aroids).

applanate.   Flattened horizontally; without verticle curves or bends.

apposition (of cell wall).   Growth of a cell wall by successive deposition of wall material, layer upon layer. Compare intussuception.

appressed.   Closely and flatly pressed against the entire length of an organ or part; with an angle of divergence of 15° or less.

appressorium, *pl.* **appresoria.**   A flattened, hyphal, pressing organ of parasitic fungi which adheres to the surface of the host prior to penetration of fine haustoria.

approximate.   Close to each other but not united.

aquaculture.   The farming of fish and other organisms in an aquatic medium.

aquatic.   Living or growing naturally in or underwater.

aquifer.   A subsurface, water-bearing stratum or layer in the earth.

arachnoid.   Cobwebby by soft and fine entangled hairs which are longer than tomentose.

arbor.   **1.** A tree. **2.** A latticework trellis covered with vines or branches for shade.

arborescent.   Of treelike habit; resembling a tree in growth or appearance.

arboretum, *pl.* **arboreta.**   A place where trees, shrubs, and other plants are grown for scientific and/or educational purposes.

arbuscula.   A small or low shrub having the form of a tree.

arch-, archeo-.   A prefix meaning ancient.

arch-, arche-, archi-.   A prefix meaning first or beginning.

archegonial.   Of or pertaining to the archegonium.

archegonial chamber.   A small cavity or groove at the apex of the female gametophyte in certain gymnosperms, especially the cycads; the archegonial neck cells are on its surface.

Archegoniatae.   A collective designation for all plants which have archegonia; includes the Bryophyta, all lower vascular plants, the ferns, and the majority of the gymnosperms.

archegoniophore.   A specialized, umbrella-shaped stalk or branch that bears the archegonia at its apex, as in certain liverworts. (*See* fig. H-2).

archegonium, *pl.* **archegonia.**   The female reproductive organ of the Archegoniatae, usually a flask-shaped structure with a swollen base or venter containing an egg, and a slender, elongated neck. After fertilization the embryo develops in the venter. (*See* fig. A-9, A-18).

archesporium.   A mass of cells in the capsule of a bryophyte from which sporogenous cells originate.

archetype.   A generalized type of structure assumed to have been possessed by ancestors of a species or group.

archicarp.   1. The initial stage of a fruiting body. 2. The female reproductive organ of the Ascomycetes (sac fungi) consisting of a filamentous portion and a basal fertile area termed the ascogonium.

Arcto-Tertiary Flora.   A temperate vegetation that occupied high latitude northern land masses around the world early in Tertiary time, and represented by the conifers and hardwoods that are now found in the forests of the Northern Hemisphere.

arcuate.   Moderately curved or arched, like a bow.

arenaceous.   Referring to those plants which grow in sand or in sandy places.

areola, *pl.* **areolae.**   A small space on the surface of a plant organ which differs in

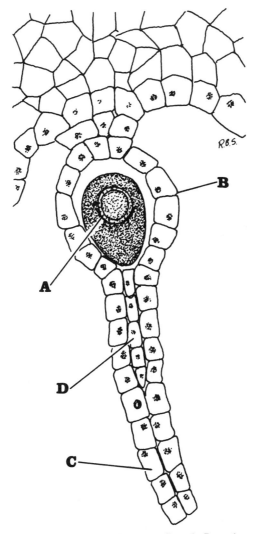

Fig. A-18. Archegonium: A. egg, B. venter, C. neck, D. neck canal cells.

texture, structure, or color from the surrounding area; often used in leaves for the area between small veins; small spinebearing areas in the cactus family, Cactaceae; in lichens, a small polygonal area of the thallus.

areolate. Divided into many angular or squarish spaces.

areole. A small area in a leaf between small veins.

argillaceous.   Clay-colored, drab; resembling or containing clay.

argillophytes.   Plants dwelling on mud, clay or on mud banks.

arhizal, arhizous.   Referring to plants without true roots, e.g. mosses.

aril.   An appendage, outgrowth, or outer covering of a seed, growing out from the hilum or funiculus; it is sometimes pulpy or fleshy as in the exterior coat of some seeds. (*See* fig. A-19).

arillate.   Having an aril.

arillode.   A false aril; a fleshy coat derived from the orifice or micropylar rim of the outer integument instead of from the stalk of an ovule, as in the mace of nutmeg.

arista.   A bristlelike appendage as on the glumes of many grasses; a bristly awn.

aristate.   Bearing a stiff bristlelike awn or seta; tapered to a narrow elongated apex.

aristulate.   Bearing a small awn.

armature.   Any covering or occurrence of spines, hooks, barbs or prickles on any part of the plant.

armed.   Having thorns, spines, prickles, etc.

armor, armored.   **1.** Being armed; a covering of old leaf-bases on the stems of cycads. **2.** Possessing a protective covering of scales or plates which cover the cell surface, as in certain dinoflagellates.

Fig. A-19. Aril.

aromatic.   1. An organic compound composed of closed rings of carbon atoms; e.g. essential oils of plants. Compare aliphatic, heterocyclic. **2.** Having an aroma.

arrangement.   Referring to the disposition of plant organs or parts with respect to each other.

arroyo.   A dry gully or watercourse.

arthrospores.   1. Spores formed from the fragmentation of fungal hypha. *Syn.* oidium. **2.** Thick-walled vegetative resting spores of certain algae.

article.   One of the segments of a jointed fruit, such as a loment.

articulate, articulation, articulating.   Jointed; having nodes or joints, which are natural points of separation at maturity.

articulated laticifer.   A laticifer composed of more than one cell; a compound laticifer. Compare nonarticulated laticifer. (*See* fig. L-2).

artifact.   A substance or structure seen in fixed preparations of cells and tissues which does not naturally exist in them but which results from post-mortem changes, cytological processing techniques, or limitations of equipment or the observer.

artificial classification, artificial system.   Terms referring to any method of classification which is based only upon a few convenient characters, e. g. flower color, habit, etc., regardless of actual affinities. Not intended to show true phylogenetic relationships.

artificial selection.   The result of the conscious breeding of plants by man for the goal of improving agricultural or horticultural crops. Compare natural selection.

arundinacious.   Having a reedlike stem, as in tall grasses.

ascending.   Rising obliquely or curving upward from near the base; usually at an angle of 40° to 60°.

ascidium.   A pitcher or flask-shaped organ or structure, as the leaves of *Sarracenia,* (pitcher plant). (*See* fig. A-20).

ascigerous.   The ascus stage of an ascomycete.

asciiform.   *See* dolabriform.

ascocarp, ascoma.   In Ascomycetes (sac fungi), a sporocarp producing asci and ascospores. There are three kinds: the apothecium, perithecium, and cleistothecium.

ascogenous hypha.   A specialized hypha which produces one or more asci in the *As-*

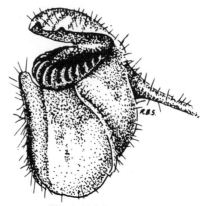

Fig. A-20. Ascidium.

comycetes. The hypha contain paired haploid male and female nuclei (the dikaryon), which develop from the ascogonium. (*See* fig. A-21).

ascogonium, *pl.* **ascogonia.**   The female gametangium in the Ascomycetes (sac fungi); the cell or group of cells which receives the nuclei from the antheridium and from which the ascogenous hyphae arise.

ascomycete.   A fungus of the class Ascomycetes (sac fungi) which produces spores in saclike cells (asci); e.g. powdery mildews, yeasts, cup fungi.

ascospore.   A haploid spore formed within an ascus following sexual fusion and meiosis.

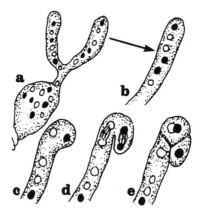

Fig. A-21. Ascogenous hyphae: a. ascogenous hyphae, b–e. shows steps in initiation of ascus formation.

ascostroma, *pl.* **ascostromata.**   A stromatic ascocarp in which the asci are produced in locules.

ascus, *pl.* **asci.**   A tubular, saclike structure characteristic of the Ascomycetes (sac fungi) in which usually eight haploid ascospores develop following karyogamy, meiosis and mitosis. (*See* fig. A-17).

ascus mother cell.   The binucleate crook cell in Ascomycetes (sac fungi) which develops into the ascus following karyogamy.

-ase.   A suffix for most all enzyme names.

asepalous.   Lacking sepals or calyx.

aseptate.   Without septa; having no cross walls or partitions. *Syn.* eseptate.

asexual, asexual reproduction.   Any reproductive process that does not involve the union of gametes or sex cells; in lower forms, reproduction by spores not associated with the sexual process.

aspect dominance.   The phenomenon of one species being particularly evident at a given time of year, usually due to flowering or foliation.

asperous.   Rough to the touch, scabrous, harsh.

aspirated pit.   In wood, a bordered pit in which the pit membrane is laterally displaced and the torus blocks the aperture. (*See* fig. B-6).

asporogenous.   Nonspore-forming.

asporogenous yeast.   A yeast which reproduces asexually by budding and does not produce ascospores.

assimilated energy.   The portion of solar energy striking the plant and transformed into chemical energy by the plant.

assimilation.   Constructive metabolism or anabolism; the process of taking in nonliving nutrient substances and converting them into protoplasm and cell walls.

association.   Defined by resolution as plant communities "of definite floristic composition, uniform physiognomy and when occurring in uniform habitat conditions." However, adherence to these criteria is not always possible and has resulted in two different concepts used by European and North American ecologists; a grouping of two or more consociations.

associes.   A seral plant community; a nonpermanent community to be replaced by another.

assortative mating.   The nonrandom sexual reproduction in which breeding tends to take place between male and female of a particular kind.

assortment.   The random distribution to the gametes, of different combinations of chromosomes.

assumentum, *pl.* **assumenta.**   The individual values of a silicle or silique, which are capsular fruits typical of the mustard family (Brassicaceae).

assurgent.   Ascending, rising; growing upward at an angle, but not erect.

astamonous.   A flower with no stamens.

asterad embryos.   A common embryo type of dicotyledons in which both the basal and terminal cells contribute to the development of the embryo.

astigmatic.   Lacking stigmas.

astomatous.   Without stoma; lacking pores in the epidermis.

astrosclereids.   A branched or stellate-shaped (ramified) cell; a type of sclereid often found in leaves and stems of xerophytic dicotyledons. (*See* fig. S-3).

astylocarpellous.   A type of carpel lacking both a style and a stipe.

astylocarpepodic.   A type of carpel without a style, but having a stipe.

astylous.   Without a style.

asymmetrical.   Lacking symmetry; irregular in shape or outline; zygomorphic.

asynapsis.   The failure of homologous chromosomes to pair during meiosis.

atactostele.   A stele consisting of vascular bundles scattered throughout the parenchyma tissue of the stem; e. g. characteristic of monocotyledons.

atavism.   The recurrence of appearance of traits possessed by a remote ancestor which usually results from a recombination of genes. Often called a "throw-back".

atelechore.   A plant with no specialized dispersal mechanism for disseminules.

atmometer.   An instrument for measuring evaporation from a moist surface.

atom.   The smallest unit of matter of an element that can exist alone and which makes up the 'building blocks' of molecules.

atomic mass unit.   A weight equal to 1.67×10^{-24} gram.

atomic number.   The number of protons in an atomic nucleus or the number of positive charges.

atomic weight.   The relative weight of an atom when compared to a representative atom (now oxygen) expressed in atomic weight units.

atomic weight unit.   Equals one-sixteenth of the mass of an oxygen atom.

ATP.   *See* adenosine triphosphate.

atrophy.   The reduction in size of an organ or an entire body.

atropous.   *See* orthotropous.

attenuate.   With a long, slender taper, more gradual than acuminate; applied to bases or apices of parts.

attractants.   The various methods and devices by which flowers lure or attract insect visitors, e.g. pollen, nectar, brood-place, etc.

attraction unit.   In pollination ecology, the unit which attracts the pollinators to the vicinity of a flower or flowers; can be an individual flower or an entire inflorescence.

atypical.   Not typical, irregular, departing from the norm.

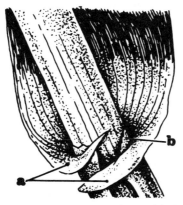

Fig. A-22. Auricle on grass leaf: a. auricle, b. ligule.

auricle. An earlike lobe at the base of various leaves and petals. (*See* fig. A-22).

auriculate. Bearing auricles.

auriculiform. Usually obovate in outline with two small rounded, basal lobes.

autecology. The study of individual species in relation to their environments.

auto-. A prefix meaning self.

autocarp. A fruit which results from self-fertilization.

autochthonous. A native species that has originated within an area and is reproducing naturally.

autocious. Having both archegonia and antheridia on the same plant but in separate clusters, as in certain mosses.

autocolony. A daughter colony produced asexually by coenobic algae.

autodeme. A group of individuals in a taxon composed of predominantly self-fertilizing (autogamous) individuals.

autoecious rusts. Referring to a parasitic rust fungus which completes its entire life cycle on the same host plant.

autogamy. Self-fertilization.

automimicry. Mimicry within a species.

automixis. Asexual reproduction where meiosis takes place, but fertilization does not occur.

autoparasite. A parasite living on a parasite.

autophene. A phenotype due to the genetic constitution of the cells of the tissue involved. Compare allophene.

autophilous. Self-pollinating, autogamous.

autophyte, autophytic. A plant containing chlorophyll that is capable of manufacturing its own food. Contrast heterotroph.

autoploid, autopolyploid. A polyploid organism which has three or more sets of chromosomes, all of which come from the same species; usually results from the doubling of chromosomes in a single individual. Compare amphiploid, alloploid.

autoradiograph. A photographic image showing the location of radioactive substances in cells or tissues, obtained by exposing a photographic emulsion to radioactive emissions.

autosome. Any chromosome other than a sex chromosome.

autospore. An asexual, nonmotile reproductive cell similar to aplanospores, and resemble in miniature, the mother cells that produce them. They are found in certain taxa of green algae, e.g. *Chlorococcum* and *Chlorella*.

autosyndesis. In polyploids and their hybrids, the pairing of chromosomes derived from the same parent.

autotetraploid. A tetraploid with four identical sets of chromosomes.

autotoxicity. Refers to the alleopathic effects between members of the same species; e.g. as between individual plants or between a plant and its propagules.

autotroph, autophyte. An organism that makes its own food.

autotrophic. Organisms which are able to manufacture all of their own food from organic compounds, as in all chlorophyll-containing plants and some bacteria.

autumnal coloration. The change in autumn of leaf color from green to reds and yellows caused by the decomposition of chlorophyll and the presence of xanthophyll (yellow), carotene (orange), and anthocyanin (red) pigments.

auxesis. Increase in size by cell enlargement, without cell division.

auxiliary cell. In some red algae (Rhodophyta), a cell to which the diploid zygote nucleus is transferred and where growth of the carposporophyte is initiated.

auxin. A natural hormone that regulates plant growth particularly through cell elongation rather than cell division. It also is important in the development of fruit and in apical dominance.

auxospore. A spore formed in certain diatoms by the fusion of two gametes.

auxotroph. A mutant microorganism requiring an external supply of some organic substances not needed by the wild type.

average. *See* mean.

average distance.   A computation which equals the square root of the area divided by density computed for each species.

awl-shaped.   Narrow and gradually tapering to a sharp point; sharp-pointed from a broader base.

awn.   A bristlelike appendage or part of, as in the back or at the tip of glumes and lemmas of many grasses.

axenic.   Germ-free; without another organism being present, as in axenic culture.

axial.   Of or pertaining to an axis, especially the main axis.

axial parenchyma.   Parenchyma cells in the axial system, in contrast with ray parenchyma cells; found in secondary xylem of dicot and conifer wood. *Syn.* vertical parenchyma.

axial system.   All cells derived from the fusiform cambial initials and oriented with their longest diameter parallel with the main axis of a stem or root and occurring in secondary vascular tissues. (*See* fig. A-23).

axial tracheid.   A tracheid in the axial system of wood, as contrasted with ray tracheids.

axil.   The angle between the stem axis and a leaf petiole, branch or other appendage attached to it.

axile.   Of, belonging to, or located in the axis; central in position.

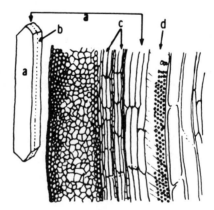

Fig. A-23. Axial system: a. fusiform initial, b. periclinal division, c. phloem, d. xylem.

axile placentation.   Placentation in which the ovules are borne at or near the center of a compound ovary, on a central axis in separate chambers in the fruit. (*See* fig. P-10).

axillary.   In an axil; a term applied to buds, branches, or meristems, which occur in the axil of a leaf; e.g. axillary buds, axillary leaves, axillary meristems. (*See* fig. L-3).

axis, *pl.* **axes.**   The main stem of a plant; the main or central line of development of any plant or organ.

azonal soils.   Soils in which pedogenesis has hardly begun and which show no obvious profile differentiation.

azonal vegetation.   An edaphic climax community controlled primarily by extreme soil conditions.

azotobacteria.   Bacteria which are able to change atmospheric nitrogen into fixed nitrogen.

azygospore.   A zygospore which develops parthenogenetically.

B

B₁, B₂, B₃, etc. First, second, third, etc. backcross generations.

baccacetum. An aggregation of berries.

baccate. Like a berry; pulpy or fleshy; producing berries.

bacciferous. Producing berries.

bacciform. Shaped like a berry.

bacillus, *pl.* **bacilli.** A rod shaped bacterium.

backcross. The crossing of a hybrid with either of its parents, or with a genetically equivalent organism.

backcross breeding. A system of breeding whereby recurrent backcrosses are made to one of the parents of a hybrid, accompanied by selection for a specific character or characters.

back mutation. A mutation of a mutant gene resulting in a reversion to its original state.

bacterium, *pl.* **bacteria.** Minute, prokaryotic organisms that usually lack chlorophyll, and exist mostly as parasites or saprophytes; a few are autotrophic. They cause decay, putrification, and fermentation, and many are pathogenic.

bactericide. A substance causing the death of bacteria.

bacteriochlorophyll. The photosynthetic pigment occurring in the purple sulfur bacteria, and the purple nonsulfur bacteria; it chemically resembles chlorophyll *a,* but its maximum absorption spectrum is in the infrared.

bacteriology. The science dealing with bacteria.

bacteriophage. A virus that parasitizes a bacterial cell. (*See* fig. B-1).

bacteriostatic agent. A substance that arrests growth of bacteria without killing them.

baculate. Pollen with pillarlike processes, always longer than broad and greater than one micrometer tall.

baculum, *pl.* **bacula.** The rod-shaped processes of baculate pollen.

48

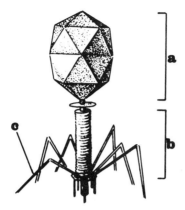

Fig. B-1. Bacteriophage: a. head, b. tail, c. tail fibers.

bagasse. A fibrous by-product of sugarcane processing used in the manufacture of various paper products.

balanced lethals. A genetically controlled system in plants in which homozygosity of nonallelic recessive lethal genes results in zygote mortality when parents are self-fertilized, thus insuring heterozygosity in the surviving progeny.

balanced polymorphism. A genetic polymorphism maintained in a population because the heterozygotes for certain alleles have a higher adaptive value than either homozygote.

balausta. A many seeded, many loculed indehiscent fruit with a thick, tough pericarp (rind); e.g. pomegranate.

bald. Without pubescence.

ballistospore. Referring to violently ejected spores.

ballochore. A plant dispersed by mechanical expulsion of seeds or fruits.

balsam ducts. The resin ducts of the conifers.

banded apotracheal parenchyma. Axial parenchyma forming concentric lines or bands in wood, usually independent of the pores, or vessels.

banner. The broad uppermost petal of a papilionaceous corolla, as in the irregular flowers of certain members of the pea family (Fabaceae); *Syn.* standard; vexillum.

barbate. Bearded with long stiff hairs (trichomes), usually in a tuft, line or zone.

barbed. Said of bristles or awns that have short terminal or lateral spinelike hooks that are bent sharply backward, as in a fishhook.

barbellate. With short, usually stiff hairs; finely barbed.

barbellulate. Finely barbed.

barbulate. Finely bearded.

bark. A nontechnical term applied to all tissues outside the vascular cambium of the xylem of woody plants. In secondary growth, bark includes the secondary phloem and periderm; thus, functional phloem is the innermost part of the living bark.

barochore. A plant dispersed by gravity by means of heavy seeds or fruit.

bars of Sanio. Thickenings of intercellular material and of the primary wall along the upper and lower margins of the pit-pairs, in conifers. *Syn.* crassula.

basal. Located at or near the base of a structure.

basal body. A cytoplasmic organelle at the base of a flagellum or cilia. It is identical in form and structure to a centriole.

basal placentation. Attachment of ovules at the base of the ovary in angiosperms; i.e. with the placenta attached to the base of the ovary. (*See* fig. P-10).

basalt. A dark grey to black colored, dense to fine-grained, igneous rock.

base. **1.** The bottom or lower portion of a body or structure. **2.** A compound that reacts with an acid to form a salt and has the tendency to acquire a hydrogen atom. Compare acid.

base analogue. A purine or pyrimidine base differing slightly in structure from a normal base that may be incorporated into nucleic acids.

base-pairing rule. The requirement that adenine pair with thymine (or uracil) and guanine with cytosine in a nucleic acid double helix.

basicaulous. Near the base of a stem.

basicidal capsule. A type of capsule that dehisces through basal slits or fissures, as in some members of the birthwort family (Aristolochiaceae).

basic number. The number of chromosomes found in the gametes of a diploid ancestor of polyploids; represented by x.

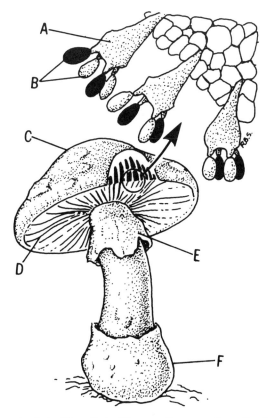

Fig. B-2. Basidiomycete: A. basidium, B. basidiospore, C. cap or pileus, D. gills, E. annulus or ring, F. volva.

basidiocarp. The sporocarp of Basidiomycetes that produce basidia and basidiospores on gills and other surfaces; the visible fruiting body of mushrooms, bracket fungi and others.

basidioles. Developing basidia.

basidiomycete. Any member of the Basidiomycetes (club fungi), a class of true fungi which includes the mushrooms, puffballs, shelf fungi, rusts, and smuts, characterized by the production of basidia and basidiospores. (*See* fig. B-2).

basidiospores. The haploid spores of Basidiomycetes produced on the basidia following karyogamy and meiosis of the diploid zygote. (*See* fig. B-2).

basidium, *pl.* **basidia.** A specialized, often club-shaped cell characteristic of the

Basidiomycetes on which the haploid basidiospores (usually four), are produced following karyogamy and meiosis. (*See* fig. B-2).

basifixed.   Referring to a structure, usually an ovule or anther, that is attached or fixed by its base to a support, rather than by its side.

basifugal development.   *See* acropetal development.

basipetal, basipetal development.   Descending; developing, or blooming in succession from an apical position toward the base; said of organs (flowers, leaves), tissues or cells which develop successively on an axis so that the oldest are at the apex. *Opp.* acropetal.

basipetiolar.   At the base of the petiole.

basiramous.   Leaves on the lower part of a branch.

basiscopic.   Toward the basal or proximal (as opposed to distal) end; facing basally.

basophyte.   Plants which normally live in basic soil, above pH 7.5.

bast bundles.   Bundles of thick-walled cells parallel to the midrib, as in the quillwort family (*Isoetes*).

bast fiber.   Any extraxylary fiber; used earlier in reference to phloem fibers. *Syn.* soft fiber.

Batesian mimicry.   A type of mimicry in which one species (the mimic) has evolved structural or behavioral characteristics that are very similar to those of another species (the model); the model has some protective feature such as bad taste or toxicity, as with some butterflies that resemble monarch butterflies, or some flies that resemble bees or hornets.

bathyal zone.   The region of oceanic water above the Continental Slope and Continental Rise.

B chromosomes.   Chromosomes that are in addition to the normal chromosome complement. They are frequently heterochromatic, although in some species they may be essentially euchromatic or may be partly heterochromatic and partly euchromatic; they may occur in varying numbers throughout the individuals of a population, or they may vary in number from one part of a plant to another; they may be smaller or larger than the normal chromosome complement. *Syn.* accessory or supernumerary chromosomes.

beak.   A firm, slender projection on certain fruits, seeds and carpels.

beaked.   Ending in a beak.

beard.   A cluster of bristlelike hairs or awns, as in some grasses; a tuft, line, or zone of pubescence, as on some corollas.

bearded.   Bearing long or stiff hairs.

beebread.   Pollen stored by bees in the honeycomb and used for food by larvae and newly-emerged workers.

beltian bodies.   Bodies produced at the tips of the leaflets of certain acacias, which produce various oils and proteins. These bodies are collected and consumed by ants (especially in the genus *Pseudomyrmex*) which live in the thorns of these plants. The ants protect the acacias from herbivorous predation.

benchland.   A relatively flat area or plateau; the tillable area of farmland.

benthonic, benthic, benthos.   Referring to plants or animals which live on or attached to the bottom of aquatic habitats.

benthophilous.   Referring to plants living submerged and attached to the bottom of seas and estuaries.

berry.   A fruit developed from an ovary containing one to several carpels in which the ovary wall (or at least its inner portions) and the inner structures of the ovary become enlarged and juicy; seeds within have their own hard seed coat; e.g. grapes, tomatoes, oranges. (*See* fig. B-3).

betacyanin.   Reddish pigments composed of complex aromatic compounds, responsible for the red color of beets, and found e.g. in the following families: goosefoot

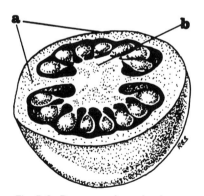

Fig. B-3. Berry: a. pericarp, b. placenta.

(Chenopodiaceae), cactus (Cactaceae), nightshade (Solanaceae), and portulaca (Portulacaceae).

beta particle.   A high-energy electron emitted from an atomic nucleus which is undergoing radioactive decay.

B-horizon.   Subsoil; one of the three common soil layers, and consisting of partially weathered soil which contains much less organic material than the A-horizon soil (humus topsoil), above it. *Syn.* B-layer.

bi-.   A prefix meaning two or twice, as in bipinnate (twice pinnate).

bias.   A consistent and false departure of a statistic from its proper value.

bibacca.   A fused double berry, as in honeysuckle (*Lonicera*).

bicarpellary, bicarpellate.   Composed of two carpels.

biciliate.   Possessing two cilia or elaters, as in motile spores.

bicollateral vascular bundle.   A bundle having the phloem on two sides of the xylem.

bicrenate.   A margin type with smaller rounded teeth on larger rounded teeth. Doubly-crenate.

bidentate.   Having two teeth.

biduous.   A plant lasting two days.

biennial.   A plant that normally requires two growing seasons to complete its life cycle. Vegetative growth occurs during the first year, with flowers and fruits produced during the second year, followed by death of the plant.

bifacial leaf.   A leaf with palisade parenchyma on one side of the blade and spongy parenchyma on the other side. A dorsiventral leaf; having a distinct upper and lower surface. Compare unifacial leaf.

biferous.   Bearing two crops of fruit in one season; appearing twice yearly.

bifid.   Forked; two-cleft; divided into two parts or lobes, as the apex of certain stigmas.

biflorous.   Flowering in autumn and spring; two-flowered.

bifoliate.   With two leaflets arising from a common point.

bifurcate.   Divided into two forks or branches; as in Y-shaped hairs, stigmas or styles.

bilabiate.   Having two lips, usually applied to a corolla or calyx; each lip may be lobed or toothed. Bilabiate corollas are common in the mint (Lamiaceae) and figwort (Scrophulariaceae) families.

bilateral.   Referring to pollen with two vertical planes of symmetry.

bilateral symmetry, bilaterally symmetrical.   Having a body composed of only two corresponding or complementary sides (or halves); symmetry in two planes, that is, where each half forms a mirror image of the other; a term often applied to flowers. *Syn.* irregular, zygomorphic. (*See* fig. B-4).

bilobed.   Having two divisions, often rounded.

bilocular.   Two-celled, or with two locules.

biloprotein.   A red and blue-green proteinaceous pigment found mainly in algae.

bimestrial.   Flowering every two months.

bimodal.   Having measurements for a given trait clustered around only two values.

binary fission.   A type of reproduction wherein a single-celled organism or cell, divides into two theoretically equal parts; e.g. bacteria.

binomial, binomial nomenclature, binomial system.   The scientific name of an or-

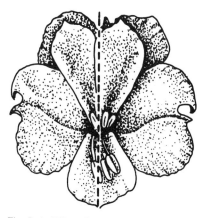

Fig. B-4. Bilateral symmetry in a flower.

ganism; a system of naming organisms in which each species is given a scientific name consisting of two words in Latin, the first designating the genus and the second, the species.

binucleate, binucleate cells.   Cells containing two separate nuclei, (dikaryotic or n+n), as those resulting from fusion of hyphae in the Basidiomycetes (club fungi).

bioassay.   A test involving the response of a living cell or organism to any artificial stimulus.

biochemistry.   The chemistry of living matter.

biochore.   That portion of the earth's surface having a life-sustaining climate.

biochrome.   A natural pigment or coloring matter in a plant or animal.

biodegradable.   Pertaining to organic substances that can be broken down relatively quickly by normal environmental processes.

biogenesis.   The doctrine that living things originate only from preexisting living things. Contrast abiogenesis.

biogeocoenosis.   An ecological concept of considering the phytocoenosis (plant community) together with its environment. It defines a specific unit of plant synecological interest.

biogeography.   The science which deals with the geographical distribution of organisms. *See* phytogeography.

biological clock.   The apparent ability of plants and animals to adjust their behavior or morphology, (e.g. development of flower buds, or breaking of seed dormancy), as the result of varying durations of daylength. *See* circadian rhythms.

biological control.   The use of parasitic, predaceous, or pathogenic organisms in the control of noxious or injurious plants or animals.

biological species.   Variously defined, but in general denotes those species or races that interbreed continuously or occasionally with each other and are reproductively isolated from other such groups. *See* hologamodeme.

bioluminescence.   The emission of light by certain living organisms.

biomass.   The amount of organic matter of a species, per unit of area or volume; the total dry weight of all organisms in a particular habitat or area; the standing crop. The term is used to express population density.

biome.   A group of similar types of communities characterized by the distinctiveness of the life-forms of the principle climax species they contain, and by their responses to that environment. Terrestrial biomes are named after their plant dominants, as pinon-juniper, tropical rain forest, desert, tundra, etc.

biometry.   The branch of science that deals with statistical procedures in biology.

biosphere.   The largest, all-encompassing ecosystem including soil, water (all of the oceans), and the atmosphere. *Syn.* ecosphere.

biosystematics.   The part of systematics that is concerned with the study of variation and evolution of a species or species complex. Biosystematic research often involves the study of breeding systems (reproductive biology) and chromosomes (cytology).

biota.   The plant and animal life of a region; a collective term for the flora and fauna.

biotic.   Pertaining to life.

biotic potential.   The total reproductive potential of an individual or a population; it is directly related to the number of functional gametes and seeds produced.

biotic selection.   Selection influenced by the biotic environment.

biotype.   A group of individuals with the same genotype; a genotypic race. Biotypes may be homozygous or heterozygous.

biovulate.   Containing two ovules.

biparous.   Having two branches or parts.

bipartite.   Divided into two parts almost to the base; two-parted.

bipinnate.   Doubly or twice pinnate; the condition of a leaf in which both primary and secondary divisions are pinnate.

bipinnatifid.   Twice pinnately cleft.

bipolar germination.   The germination of a spore by the production of a germ tube at each end.

bipolarity.   **1.** A condition of sexual compatibility in certain Basidiomycetes (club fungi) in which two basidiospores of each basidium are of one strain, and two are of another. **2.** Used for plants occurring on both north and south poles but not between.

bisected.   Completely divided into two parts.

biseriate. In two whorls or cycles; in two rows or series, as a perianth composed of a calyx and a corolla.

biseriate ray. A ray two cells wide found in secondary vascular tissues of dicots, and infrequently in conifer wood.

biserrate. A margin type with smaller sharply cut teeth on the margins of larger sharply cut teeth. Doubly-serrate.

bisexual. Having both male and female sexual reproductive structures on one individual, or occurring and functional in one flower; hermaphroditic.

bisporic embryo sac. One of three major types of embryo sac development in which an embryo sac is formed from one of the two dyad cells formed after division I of meiosis. Compare monosporic and tetrasporic embryo sac.

bitunicate ascus. An ascus with two walls, the inner elastic and expanding greatly beyond the rigid, outer wall, at or before the time of spore liberation.

bivalent. The associated (synapsed) pair of homologous chromosomes, one pair from each parent, which are observed at prophase I of meiosis; each bivalent is composed of four chromatids. (*See* fig. B-5).

bivalve. Having two valves, as in the diatoms.

bladder. 1. An expanded, often gas-filled chamber or structure. 2. A specially modified insectivorous leaf, tubular or pitcher-shaped, used to capture insects in the bladderwort family (Lentibulariaceae).

bladdery. Expanded or inflated with very thin walls.

blade. 1. The flat expanded portion of a leaf, petal or other structure. 2. The flat, leaflike thallus of certain brown algae, as the kelps. *Syn.* lamina. (*See* figs. H-4, L-3).

blastocarpous. The germination of seeds while within the pericarp (ovary wall).

blastospore. An asexual spore produced by budding.

B-layer. *See* B-horizon.

blepharoplast. A cytoplasmic granule from which a flagellum originates.

blight. A common name for any plant disease where there is a sudden wilting and/or death of leaves, stems, flowers, or entire plants.

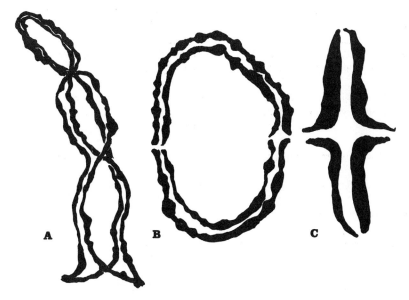

Fig. B-5. Bivalent; Various configurations of bivalents at prophase I of meiosis: A. typical, B. ring or loop, C. cross-shaped.

blind pit.   A pit without a complementary pit in an adjacent wall, which may face the lumen of a cell or an intercellular space.

block inheritance.   Inheritance of a group of genes as a unit held together by small segmental arrangements of the chromosomes.

block meristem.   *See* mass meristem.

bloom.   **1.** The state of flowering; to yield or produce flowers. **2.** The whitish waxy or powdery coating of the fruit or leaves of certain plants, e.g. plums or grapes. **3.** The sporadic seasonal occurrence of enormous numbers of algae in fresh and marine waters; e.g. as caused by various blue-greens, diatoms and dinoflagellates.

blossom.   The flower of an angiosperm; to bloom or put forth blossoms.

blotched.   Referring to the distribution of color on a plant, especially on the leaves, in broad irregular blotches.

blue-greens.   Unicellular, colonial or filamentous prokaryotic organisms of the phylum Monera. They occur in terrestrial, subterranean, and aquatic (marine and fresh water) habitats. These organisms were formerly classified with the true algae (now phylum Protista) but are more closely related to bacteria. *Syn.* blue-green algae.

B.O.D.   An abbreviation meaning "biological oxygen demand" in reference to the oxygen required for the survival of aquatic organisms.

body cell.   In gymnosperms, a cell of the male gametophyte (pollen grain), formed when the generative cell divides mitotically producing it and the stalk cell. The body cell eventually divides into sperms (two or more depending on the taxa).

bog.   Wet spongy earth, composed primarily of decayed and decaying vegetable matter.

bole.   The main trunk or stem of a tree; a strong unbranched caudex.

boll.   The pod or capsule of a plant, as a cotton boll.

bolting.   The abnormal formation of stems on plants which normally have a rosette growth habit, in response to application of growth hormones, e.g. gibberellins.

bond.   Chemical force which holds atoms together forming molecules.

bonsai.   The culture of miniature potted trees, which are dwarfed by stem and root pruning and controlled nutrition. Also, a tree or shrub produced by such treatment.

boraginaceous hairs.   Unicellular to multicellular, conical, calcified and/or silicified bristles.

Bordeaux mixture.   A mixture of copper sulfate, lime and water used agriculturally as a fungicide to control downy mildew and other fungi.

bordered.   Having a margin that is different from the rest of the structure. (*See* fig. B-6).

bordered parenchyma.   The bundle-sheath parenchyma of dicots.

bordered pit.   Pits in which the secondary wall arches over the pit membrane; typical of gymnosperm tracheids. Compare simple pit. (*See* fig. B-6).

bordered pit-pair.   An intercellular pairing of two bordered pits.

boreal.   Northern.

boreal forests.   Coniferous forests of the northern hemispheres characterized by evergreen conifers such as spruce, fir, pine.

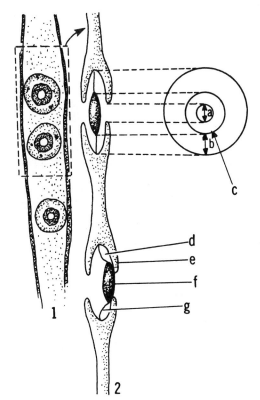

Fig. B-6. Bordered pits: 1: face view, 2: longitudinal section; a. aperture, b. margo, c. torus, d. pit cavity, e. border, f. torus of aspirated pit, g. pit membrane.

boss. *See* umbo.

bostryx. A helicoid cyme.

botanize. To study plants in the field or to collect plants for botanical purposes.

botany. The science which deals with the study of plants and plant life.

botryoid. Having the shape of a bunch of grapes; with many rounded protuberances.

botuliform. Sausage-shaped.

botulism. Food poisoning caused by the toxin of the anaerobic bacterium *Clostridium botulinum.*

boundary apotracheal parenchyma. A type of apotracheal parenchyma which occurs in dicot wood. The axial parenchyma cells either occur singly or form a more or less continuous layer at the beginning of a season's growth (initial), or at the close of the season (terminal). *Syn.* terminal apotracheal parenchyma.

boundary layer. A layer of air in contact with a leaf and influenced by the shape of the leaf. The layer is thinner for small leaves and becomes thinner with increasing wind velocities.

brachysclereid. A short, roughly isodiametric sclereid, resembling a parenchyma cell in shape. Stone cells; the sclereids of barks and fruits. (*See* fig. S-3).

bracket fungus. A saprophytic basidiomycete fungus whose basidiocarps (sporophores), grow horizontally on trees forming shelflike structures (called brackets or conks). (*See* fig. B-7).

brackish. Referring to habitats which have salinity less than a normal marine environment; salinity usually less than two percent.

bract. 1. A modified, often much reduced leaf subtending a flower or inflorescence; morphologically a foliar organ, e.g. the red bracts of *Poinsettia.* **2.** In conifers, a modified scalelike leaf of a pine cone (strobilus). **3.** A phyllary.

bracteal nectary. A type of extrafloral nectary found on glands.

bracteate. Possessing or bearing bracts.

bracteolate, bracteole. Possessing small bracts; a secondary bract, often very small.

Fig. B-7. Bracket fungi.

bracteose.   Having many bracts.

bractlet.   A small secondary bract borne on a pedicel or petiole, instead of subtending it.

brambles.   Any prickly shrub or vine, as raspberries or blackberries in the genus *Rubus* (Rosaceae).

bran.   The outer layers of a grain fruit (caryopsis), that are removed in milling.

branch.   **1.** To produce growth of side shoots. **2.** A shoot or secondary stem which grows from an axillary bud on the main stem. (*See* fig. B-8).

branch gap.   An area of parenchyma in the vascular cylinder of the main stem, where the branch traces are bent away from the vascular region of the main stem toward the branch. This is best seen in cross sections of the nodal region of a stem.

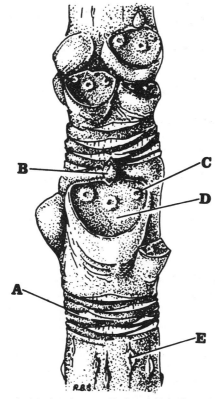

Fig. B-8. Branch: A. terminal bud scale scars, B. lateral bud, C. vascular bundle scar, D. leaf scar, E. lenticel.

branching pollen grain.   *See* polysiphonous.

branch root.   A root which arises from an older root; also called lateral or secondary root if the older root is the primary or taproot.

branch trace.   A vascular bundle in the main stem which extends into a lateral branch and becomes the vascular system for that branch. A branch trace is actually a leaf trace of one of the first leaves (prophylls) on the branch.

breeder seed.   Seed produced by an agency which is sponsoring a variety, and is used to produce foundation seed.

breeding.   The art and science of changing plants or animals genetically.

breeding systems.   Referring to the various reproductive strategies, both sexual and asexual, that occur in plants insuring survival of their species. Three different sexual systems are recognized based on the degree of inbreeding versus the degree of outbreeding: predominant outcrossing, predominant selfing, and mixed selfing and outcrossing.

bristle.   A stiff, strong trichome as in the perianth of some sedges (Cyperaceae).

bristle-cone pines.   The oldest living trees (*Pinus longaeva*), some of which have reached an age of 4900 years. The oldest are found in the White Mountains of California.

bristly.   Bearing stiff strong trichomes (hairs), or bristles.

brochus, *pl.* **brochi.**   The meshes of a reticulum of a pollen grain.

bromelain.   A protease found in the fresh juice of the pineapple.

brush hairs.   Hairs branched with ray cells radiating in all directions; e.g. the hairs in Asteraceae found on the upper part of the style.

bryology.   The study of bryophytes.

bryophyte.   Any plant of the division Bryophyta which includes the mosses, liverworts, and hornworts. Characteristics include having a dominant gametophyte generation, lacking true stems, leaves, roots and vascular tissue, and having motile sperms which require water for fertilization.

bud.   1. An undeveloped shoot containing the embryonic meristems which develop into flowers, stems, or leaves, and are enclosed in protective specialized leaves termed bud scales. Buds are often classified according to function or position, e.g. accessory, lateral, axilliary. **2.** The vegetative outgrowth or protrusion, of a bacterium or yeast cell, capable of developing into a new organism.

budding. **1.** The vegetative (asexual) reproduction of plants by grafting a stem bud of a desired species onto the root stock of another plant, often a different species or variety. **2.** A type of asexual reproduction in which a small protuberance or outgrowth (bud) develops, and becomes separated from the parent cell; as in some yeasts and bacteria. **3.** The first sign of growth in the spring, as the "trees are budding".

bud mutation, bud sport. A genetic mutation in the somatic tissue of a bud that causes it to give rise to a branch, flower or fruit that differs from other parts of the plant. Many fruits of considerable economic importance have arisen in this manner, e.g. seedless oranges and grapes, and many varieties of apples.

bud primordium. Meristematic tissue that gives rise to a lateral bud.

bud scales. Specialized protective leaves which cover the shoot apex and embryonic leaves of a winter bud, preventing dessication and injury.

bud scale scars. Scars left on a twig by the abcission of bud scales from the terminal buds of the previous year.

bulb. A short, usually globose underground stem, covered by fleshy overlapping leaf bases or scales which function as storage organs. They are found in many monocots, e.g. onion, lily, narcissis, tulip. (*See* fig. B-9).

bulbel. Small daughter bulbs arising around the mother bulb.

bulbiferous. Producing bulbs.

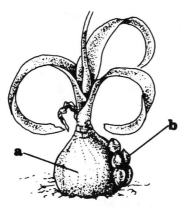

Fig. B-9. Bulb: a. bulb, b. bulbet.

bulbil, bulblet.   A small secondary bulb that develops in a leaf axil, inflorescence, or other unusual location. (*See* fig. B-9).

bulbose, bulbous.   **1.** Having the structure of or resembling a bulb. **2.** Producing or containing bulbs.

bulk breeding.   The growing of genetically diverse populations of self-pollinated crops in a bulk plot with or without mass selection, followed by single-plant selection.

bullate, bullose.   Having a blistered, swollen or puckered appearance, as the surface of certain leaves, e.g. Savoy cabbage; covered with rounded projections resembling unbroken blisters.

bulliform cells.   Large, thin-walled cells present in the leaves of most monocots. They may function in the unrolling of developing leaves, in the opening and closing movements of mature leaves, or in water storage. *Syn.* motor cell.

bunch grasses.   Referring to those grasses which do not spread by rhizomes or stolons, but which form more or less well defined clumps. Compare sod grasses.

bundle.   A group of conducting elements, the xylem and phloem, in the stem of a vascular plant.

bundle cap.   Sclerenchyma or thick-walled parenchyma associated with a vascular bundle and appearing like a cap on the phloem or xylem side, as seen in cross section.

bundle scar.   Marks on a leaf scar formed by the breaking of the vascular bundles (leaf traces) that pass from the twig into the petiole.

bundle sheath.   A special layer, or layers of cells, which enclose the vascular bundles of dicot and monocot leaves. The cells are often parenchyma, but some are sclerenchyma. *Syn.* vascular bundle sheath. (*See* figs. C-8, M-4).

bundle sheath extensions.   Somewhat thick-walled cells of dicots which extend from the bundle sheath toward the epidermis, some terminating in the mesophyll, others reaching the epidermis. The cells may consist of parenchyma or sclerenchyma, and apparently function in conduction.

bundle tip.   The end of a vascular bundle in a leaf, usually a single vessel or tracheid.

bunt.   A disease of wheat and grasses, caused by various species of the smut fungus *Tilletia*.

bur.   A rough and prickly covering of a fruit, usually functioning as a dispersal mechanism; as in the cocklebur (*Xanthium*) or bedstraw (*Galium*).

burl. **1.** An abnormal outgrowth or excrescence on a tree trunk, usually due to stimulation of the cambium by insects. **2.** The enlarged root crowns of many mediterranean and chaparral perennial shrubs.

bush. **1.** A low and thick shrub, without a distinct trunk. **2.** Uncultivated land characterized by scrubby vegetation, as in Australia.

button. An immature mushroom before the expansion of the cap; pertaining to any number of structures which resemble a button, e.g. a rose hip, a bud, etc.

buttress. A type of root forming boardlike growths at the base of certain trees (e.g. bald cypress), which probably function in the upright support of the tree.

buzz pollination. *See* vibratile pollination.

C

caducous. Falling or dropping off very early, usually in reference to floral organs (e.g. petals, sepals).

caecum. An elongation or outgrowth of the embryo sac into the nucellus of certain plants, e.g. *Allium*.

caeoma. A spore-producing structure in rust fungi.

caespitose. *See* cespitose.

calathiform. Cup-shaped.

calcarate. With a spur; spurred.

calcareous. Containing lime; e.g. calcium carbonate, calcite, or magnesium carbonate.

calceiform, calceolate. Shoe or slipper-shaped.

calcicole, calcicolous, calciphilous. Plants which live in chalky or limestone soils.

caliciform. *See* calyciform.

calcifuge. A plant which does not grow well in calcareous soils.

callose. A complex carbohydrate found in seed plants which is a common wall constituent associated with the sieve areas of sieve elements; it may also develop in reaction to injury in sieve elements and parenchyma cells.

callosity. On bark, a hardened or thickened area.

callus. **1.** A tissue composed of large thin-walled parenchyma cells which develop on or below a wounded surface often resulting in a firm thickening or protuberance. **2.** Undifferentiated tissue; a term used in cell tissue culture. **3.** In grasses, the thickened extension at the base of the lemma at the point of its attachment to the rachilla.

calorie. The amount of heat energy required to raise the temperature of one gram of water one degree centigrade (1°C). Most commonly used in metabolic studies is the kilocalorie (Kcal) or Calorie, which is the amount of heat required to raise the temperature of one kilogram of water one degree centigrade.

Calvin cycle. The incorporation of carbon dioxide into glucose through a cyclic series of enzymic reactions.

calvitium. An area lacking hairs; a bald spot.

calybium. A hard one-loculed, dry fruit derived from an inferior ovary, as in *Quercus* (oaks).

calycanthemy. The abnormal development of colored petallike structures in place of a normal calyx.

calyciflorous. A condition where stamens and petals are adnate to the calyx.

calyciform, calyculate. Calyxlike; bearing a set of small bracts beneath or against the calyx which resemble an extra or outer calyx; e.g. the short bracts of an outer involucre of certain members of the sunflower family (Asteraceae).

calyculus. A simulated calyx composed of bracts or bractlets.

calyptra. A hood or cap; the enlarged haploid archegonium that surrounds and protects the developing embryonic sporophyte of mosses, liverworts, and some lower vascular plants, (*See* fig. C-1); also, the lid of eucalyptus fruit.

calyptrogen. The apical meristem in roots which gives rise to the rootcap independently of the initials of the cortex and central cylinder (stele).

calyx, *pl.* calyces. A collective term for all the sepals of a flower; the lowermost whorl of floral organs.

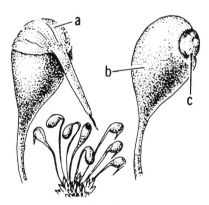

Fig. C-1. Calyptra: a. calyptra, b. capsule, c. operculum.

calyx-lobe. A free projecting part of a gamosepalous calyx.

calyx tube. The tube of a gamosepalous calyx.

cambial initials. Cells of the vascular cambium which give rise to two kinds of cells: the fusiform initials (source of axial cells of xylem and phloem) and ray initials (source of the ray cells).

cambial zone. A region of thin-walled, meristematic cells between the secondary xylem and secondary phloem. Composed of cambial initials and their recent derivatives that have not yet differentiated into mature cells.

cambium. Lateral meristem in vascular plants which produces secondary xylem, secondary phloem, and parenchyma, usually in radial rows; it consists of one layer of initials and their undifferentiated derivatives. *See* cork cambium, vascular cambium.

campanulate. Bell-shaped; with a flaring tube about as broad as long and with a flaring limb or lobes.

camptodromous. A type of leaf venation in which the secondary veins curve towards but do not terminate at the margins.

campylodromous. Said of leaf venation with several primary veins or their branches diverging from a single point and forming basally recurved arches which converge toward the apex.

campylotropous. Said of an ovule or seed which is curved on one side so that the micropyle (apex) and the funiculus (stalk) are approximately at right angles to each other; an ovule or seed curved to bring the micropyle and funiculus close together.

canaliculate. Channeled or grooved longitudinally (lengthwise), usually in reference to petioles or midribs.

canalization. A homeostatic process which favors a particular morphological or physiological developmental pattern or pathway, which once initiated, results in achieving a standard or normal phenotype; this occurs despite genetic or environmental disturbances during the development. The degree of canalization is under genetic control, and is typically more pronounced in animals than in plants.

cancellate, clathrate. Latticed; resembling latticework.

candelabra hairs. Stellate hairs in two or more tiers.

canescence, canescent. Covered with dense, fine, grayish-white hairs; becoming hoary, usually with a gray pubescence.

canker. Various fungal diseases of plants, especially trees and shrubs, characterized by presence of decay and dead bark surrounded by swollen margins.

canopy.   The cover or horizontal projection of the vegetation of a plant formed by its leaves, branches, etc.

cantharophilous, cantharophily.   Pollination by beetles.

cap.   The head, or pileus of a mushroom; the removable covering of a part, as the calyptra of a moss capsule. (*See* fig. B-2).

cap cells.   Cells which develop at the ends of the suspensor cells, in developing embryos of gymnosperms and some cycads.

capillary, capillate.   Resembling a hair; very slender.

capillary soil water.   Water held in the soil against the force of gravity by capillary force, e. g. adhesion and cohesion; the total amount of water in the soil available to a plant at any given time.

capillitium, *pl.* capillitia.   Sterile, threadlike structures forming a delicate, branched network among the spores in the fruiting bodies of many Myxomycetes (slime molds) and Gasteromycetes (puffballs). (*See* fig. C-9).

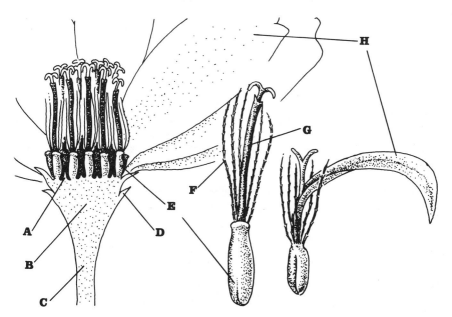

Fig. C-2. Capitulum—longitudinal section of composite flower: A. chaffy bract, B. receptacle, C. peduncle, D. phyllaries, E. inferior ovary, F. pappus, G. disk flower, H. ray flowers.

capitate.   Formed like a head; aggregated into very dense clusters or heads, especially in reference to flowers.

capitulum.   A head; a dense infloresence composed of a determinate or indeterminate group of sessile or subsessile flowers on a compound receptacle. (*See* fig. C-2).

caprification.   A method of artifically pollinating cultivated figs, by placing fruits of the wild fig (caprifig), containing fig wasps, in the trees of cultivated figs.

capsomere.   One of the protein units from which the shell of a virus is constructed; e.g. one of 2200 identical ovoid protein units which comprise the protein coat of tobacco mosaic virus (TMV).

capsule.   **1.** A dry, dehiscent fruit derived from a compound ovary of two or more carpels; also, a dry indehiscent fruit from a two or more loculed ovary. **2.** The protective colloidal sheath surrounding the cells of certain bacteria and algae. **3.** The sporangium of some bryophytes, e.g. liverworts and mosses. (*See* fig. C-1).

carbohydrates.   Organic compounds composed of carbon, hydrogen and oxygen, with a ratio of hydrogen to oxygen of 2:1; e.g. sugar, cellulose, starch.

carbon cycle.   **1.** The cyclical pathway of carbon atoms during photosynthesis and respiration. **2.** The circulation and utilization of carbon atoms through the biosphere.

Carboniferous period.   A period during the Paleozoic era characterized by the occurrence of forests of large plants which are the source of coal deposits. *Syn.* Coal age.

cardinal.   Red in color.

cardinal temperatures.   The three temperatures which affect the growth of plants, and vary from species to species; e.g. the minimum, optimum and maximum temperatures.

carina.   A keel; the two lower petals of a papilionaceous flower; a longitudinal projection on the lower surface of the glumes and lemmas of many grasses.

carinal canal.   Small canals which form a ring around the central canal in stems of horsetails (*Equisetum*), and probably function in water conduction. (*See* fig. C-4).

carnivorous plants.   Plants that are able to utilize proteins and minerals (especially nitrogen and phosphorous), obtained from trapped animals (mostly insects). They are divided into active trappers (e.g. *Dionaea, Utricularia*), and passive trappers (e.g. *Nepenthes, Drosera, Sarracenia*).

carnose.   Fleshy.

carotene. A general name for a group of orange or yellow hydrocarbon carotenoid pigments, synthesized by plants and found in chloroplasts and chromoplasts.

carotenoids. A class of fat-soluble pigments that includes the carotenes (yellows and oranges) and the xanthophylls (yellow); found in chloroplasts and chromoplasts of plants.

carpel. In angiosperms, the ovule-bearing structure (megasporophyll) of a flower, often regarded as a single, modified, seed-bearing leaf. Usually consists of a swollen basal portion (ovary), an elongation of the ovary (style), and a receptive tip (stigma). One, several, or many carpels may occur in a flower and they may be separate from each other or fused together. Collectively, the carpels are called the gynoecium. *Syn.* pistil.

carpellate. Possessing, bearing or comprised of carpels.

carpogonium. The female gametangium (sex organ) of certain red algae (Rhodophyta), e.g. *Polysiphonia*. The enlarged basal portion contains a nucleus which functions as an egg.

carpophore. In the umbel family (Apiaceae), a thin wiry stalk that supports each half (carpel) of the pendulous dehiscent fruit; an extension of the floral axis between adjacent carpels.

carpopodium. A short, thick, pistillate stalk.

carposporangium, *pl.* **carposporangia.** In red algae (Rhodophyta), the sporangium in which the carpospores (2n) are produced. The carpospores are enclosed in a cystocarp in certain taxa of red algae.

carpospore. In red algae, a uninucleate nonmobile spore produced within a carposporangium; depending on origin it may be haploid or diploid.

carposporophyte. In certain red algae (Rhodophyta), a group of diploid cells composed of carposporangia and carpospores, plus their sterile covering cells (the pericarp); this structure is sometimes considered to be a separate plant because it is a diploid structure existing epiphytically on the female gametophyte.

carpotaxis. Referring to the arrangement of reproductive fruits on an axis.

carrageenin. Commercially valuable polysaccharides occurring in the cell wall of certain red algae; often used as an emulsifier.

cartilaginous. Like cartilage in texture; firm and elastic.

caruncle, carunculate. An outgrowth from the integuments at or near the hilum of certain seeds, e. g. castor bean (Euphorbiaceae).

caryopsis. A one-seeded, dry, indehiscent fruit with the seed coat adnate to the fruit wall (pericarp), derived from a one-loculed superior ovary; the grain or fruit of grasses, e.g. wheat, corn.

casparian strip. A bandlike portion of the primary walls of some vascular plant cells, containing lignin and suberin. They are common in endodermal cells of roots, and are found in radial and transverse anticlinal walls. Forces water entering the roots to pass through a cell membrance. (*See* fig. C-3).

cast. A fossil that results from mineralization of plant organs in which the organic material has been entirely replaced by rock.

castaneous. Chestnut-colored; dark brown.

catabolism. The degradative aspects of metabolism. Includes processes involved in the breakdown of complex substances into simpler ones, especially involving the release of energy. *Opp.* anabolism.

catadromous. Having the first lobe or segment of a pinna arising towards the apex in compound leaves.

catalyst. Any substance that accelerates the rate of chemical reaction without being used up by that reaction, e.g. enzymes.

cataphyll. Small scale leaves of the rhizomes of monocots and dicots; protective winter buds of shrubs and trees; scale leaves of cycad seedlings. They function in protection and/or storage. *See* hypsophylls.

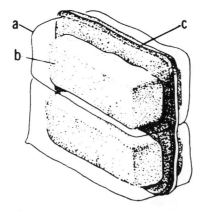

Fig. C-3. Casparian strip: a. endodermis, b. protoplast, c. casparian strip.

cation. A positively charged ion. Contrast anion.

catkin. An elongate, pendulous, deciduous cluster (spike) of unisexual, apetalous, and usually bracteate flowers, found only in woody plants, e.g. willows, oaks; an ament.

caudate. Bearing a tail, or taillike appendage.

caudex. A short, thickened, often woody, verticle or branched perennial stem, usually subterranean or at ground level.

caudicle. The strap-shaped stalk that connects a mass of pollen (pollinium) to the stigma in many orchids.

caulescent. With an obvious leafy stem; having an evident stem above ground. Contrast acaulescent.

caulid, caulidium. The main shoot, or central axis of a leafy bryophyte gametophyte which supports the vegetative and reproductive organs.

cauline. Pertaining or belonging to an obvious stem or branch; growing on a stem, as in cauline leaves.

caulis. The stalk or stem of a plant.

caulocarpic. Plants with stems living for many years, bearing flowers and fruits.

caulous. With branches more or less evenly spaced along a trunk.

cavitation. The formation of gas bubbles of various dissolved gases, in columns of water; e.g. as in columns of water within the xylem.

cell. 1. The locule of an ovary containing the ovules; a cavity (theca) of an anther containing the pollen. 2. A unit of plant structure consisting of cell wall and protoplast.

cell division. The division of the cytoplasm by the formation of a cell plate between two nuclei, resulting in two daughter cells. *Syn.* cytokinesis.

cell membrane. *See* plasma membrane.

cellobiose. A disaccharide resulting from partial chemical hydrolysis of cellulose.

cell plate. A structure that forms across the equator of the spindle in dividing cells of plants and in a few green algae during early telophase; the predecessor of the middle lamella.

cell sap.   Collective term for the fluid content of the vacuole.

cell tissue culture.   *See* tissue culture.

cellular.   Of or pertaining to cells, as cellular respiration; composed of or derived from cells.

cellular slime mold.   A common name for a fungus in the class Acrasiomycetes; a group of organisms which superficially resemble the true slime molds (class Myxomycetes), but differ in many respects.

cellulase.   An enzyme that hydrolyzes cellulose, yielding glucose; produced by many decay causing bacteria and fungi, and by some wood-eating insects and mollusks; it is also produced by microorganisms (protozoans) in the digestive tracts of certain insects (e.g. termites) and herbivorous mammals (e.g. ruminants; cattle, goats, sheep, etc.).

cellulose.   A complex carbohydrate, the chief component of the cell walls of most plants; it consists of long chainlike molecules of glucose which form microfibrils. (*See* fig. M-5).

cell wall.   A rigid layer or layers enclosing the protoplast of a plant cell, composed primarily of cellulose. *See* primary cell wall and secondary cell wall.

cenospecies.   1. Ecospecies related such that they may exchange genes among themselves to a limited extent through hybridization; usually the equivalent of biological species. 2. A group of species which when hybridized, produce partially fertile hybrids.

Cenozoic era.   The most recent geologic era of earth history, beginning 65 million years ago, during which a great expansion of the flowering plants and mammals occurred; often called the "Age of Mammals".

central canal.   The central cavity in aerial stems of horsetails (*Equisetum*), formed by a breakdown of the pith. (*See* fig. C-4).

central cylinder.   A general term referring to the vascular tissues and associated ground tissue in stems or roots. *See* stele.

centralium.   A central lengthwise cavity found in the seeds of some palms.

central mother cells.   Large, vacuolated, subepidermal cells in the apical meristem of certain gymnosperms; these cells are derived from surface initials.

central nodule.   A thickened region in the center of the valves of motile, pennate diatoms.

centric diatom.   A radially symmetrical diatom, in valve view. (*See* fig. D-2).

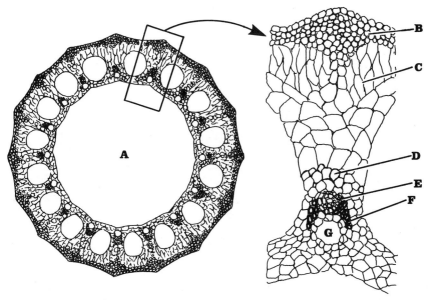

Fig. C-4. Central canal: A. central canal, B. epidermis, C. cortex, D. endodermis, E. phloem, F. xylem, G. carinal canal.

centric mesophyll.   A type of isobilateral mesophyll in which the upper and lower palisade layers form a continuous layer. Often found in narrow or cylindrical leaves.

centrifugal.   In inflorescences, blooming from the inside outward, or from the apex to the base.

centriole.   A cytoplasmic organelle found in certain flagellated plant cells, usually outside the nuclear envelope; it doubles prior to mitosis and the two resulting centrioles move to opposite poles to organize the spindle apparatus. *Syn.* centrosome.

centripetal.   In inflorescences, blooming from the outside inward, or from the base upward.

centromere.   That region along a chromosome to which some of the spindle microtubules become attached. *Syn.* kinetochore.

centroplasm.   The inner, colorless region (central body) of a blue-green algal cell.

centrosome.   *See* centriole.

centrum.   **1.** The large central air space in hollow stems. **2.** The structures enclosed by the ascocarp wall in the Ascomycetes (sac fungi).

ceraceous, ceriferous.   Waxy; wax-bearing.

cercidium.   A gall produced by fungi or insects as a result of infection; an abnormal plant growth.

cereal.   Any grass whose fruits serve as food; (from Ceres, the goddess of agriculture).

cernuous.   Nodding or drooping.

certation.   The differential growth rate of different genotypes of pollen or pollen tubes, which affects their chances of accomplishing fertilization; competition in the growth of pollen tubes down the style.

certified seed.   Seed used for commercial crop production which meets rigid standards of purity and germination and is produced by an authorized agency.

cespitose, caespitose.   Matted, growing in tufts or small dense clumps; plants forming a cushion. (*See* fig. C-5).

chaff.   **1.** The floral parts (e.g. palea, lemma) of the seeds of various cereal grains (grasses) which are removed during milling. **2.** Small bracts on the receptacle of flowers of various composites (Asteraceae). (*See* fig. C-2).

chalaza.   The basal region of the ovule where it joins the funiculus. The region of a seed opposite the micropyle.

chalazogamy.   Entry of the pollen tube through the chalazal end of the ovule. Compare porogamy.

chamaephyte.   Perennials, herbs or low shrubs, with buds less than a half meter from the ground. Compare phanerophyte.

channeled.   Marked with one or more deep longitudinal grooves.

chaparral.   Xerophytic vegetation characterized by low, mostly small-leaved, evergreen shrubs or small trees, which form dense, often impenetrable thickets.

Fig. C-5. Cespitose.

character. An attribute of an organism resulting from the interaction of a gene or genes with environment; i. e. the phenotype of an organism that results from the interaction of its genotype with the environment.

character displacement. The process of genetic divergence of two or more species when they come into contact; results from the harmful interaction of certain individuals of one taxon with individuals of another taxon (usually species). Progeny of interacting individuals are differentially eliminated because of reduced fitness, normally resulting in an increased phenotypic (character) difference between the species where their ranges overlap.

chartaceous. Papery in texture, opaque and thin.

chasmogamous. Referring to pollination which takes place in open flowers. Compare cleistogamous.

chasmogamy. The opening of the perianth at flowering time.

chasmophilous. Dwelling in rock crevices.

cheiropterophily. See chiropterophily.

chemiosmotic theory. Energy from light or oxidation drives protons across a membrane where the energy-rich compound ATP is formed as the protons flow back through a complex of enzymes.

chemoautotroph. Autotrophic organisms that derive energy from chemical reactions such as oxidation-reduction reactions of inorganic compounds, instead of from light (photosynthesis); e.g. *Nitrosomonas* and *Nitrobacter*.

chemosynthesis. The chemical reactions involved in the conversion of inorganic compounds (e.g. ammonium salts and nitrates), into a useable energy source of chemoautotrophic organisms.

chemotaxis. The movement of cells or organisms toward or away from a chemical stimulus.

chemotaxonomy. The use of chemical evidence including both primary and secondary metabolites, in elucidating phylogenetic relationships among groups of plants.

chernozems. Soils typical of central continental grassland areas, usually characterized by dark-colored soils of high organic content, and having a high level of biological activity; e.g. prairies of N. America, pampas of S. America, grasslands of S. Africa, Australia, and Asia.

chiasma, *pl.* **chiasmata.**   The X-shaped figures formed during crossing-over from the meeting of two nonsister chromatids of homologous chromosomes. *See* crossing-over.

chimera.   A plant composed of two or more genetically distinct tissues growing adjacent to each other; they may originate naturally by spontaneous mutation or may be induced artificially by application of certain chemicals, e. g. colchicine.

chionad.   A plant inhabiting snow-covered areas.

chiropterophily, cheiropterophily.   Pollination by bats.

chitin.   A tough, nitrogen containing polysaccharide which is a component of the cell walls of certain fungi, and also forms the exoskeleton of arthropods.

chlamydospore.   An asexual, thick-walled resistant spore produced from a fungal hyphal cell; may become separated from the hyphae and act as a resting spore.

chledophilous.   A plant living in waste places or dumps.

chloranthous.   Having green, usually inconspicuous flowers.

chlorenchyma.   General term applied to parenchyma cells that contain chloroplasts.

chlorobium chlorophyll.   The photosynthetic pigment occurring in the green sulfur bacteria; it chemically differs in several ways from chlorophyll *a*.

chlorophyll.   The green pigments found in the thylakoids of chloroplasts, which are essential for the utilization of light energy in photosynthesis.

chlorophyll *a*.   A type of chlorophyll found in a very large number of photosynthetic organisms; e.g. occurs in all photosynthetic eukaryotes and in prokaryotic blue-green algae.

chlorophyll *b*.   An accessory chlorophyll pigment that extends the light absorption spectrum in photosynthesis; occurs together with chlorophyll *a* in the vascular plants, bryophytes, green algae, and euglenoid algae.

chloroplast.   A cellular organelle of photosynthetic eukaryotes that contains chlorophyll; the site of photosynthesis. (*See* fig. C-6).

chlorosis.   Yellowing caused by loss of or reduced development of chlorophyll, as the result of extreme temperatures, lack of water, infection, iron deficiency of certain plants grown in soil of high pH, etc.

chondriome.   A collective term for mitochondria and plastids.

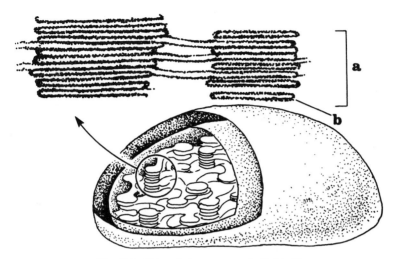

Fig. C-6. Chloroplast: a. granum, b. thylakoid.

choripetalous. Having separate and distinct petals; polypetalous.

chorology. The geographic study of the distribution of organisms.

chromatid. One of two daughter chromosomes formed by chromosome duplication, which remain attached to each other by a single centromere; they are formed during interphase (except the centromere), and become visible during prophase and metaphase of mitosis, and between prophase I and anaphase II of meiosis. *See* chromosome.

chromatin. Material contained within the nucleus of a eukaryotic cell, which is composed of DNA, RNA, and proteins (nucleoproteins); it is usually dispersed into fibers throughout the nucleus except at cell division when it condenses into chromosomes. Chromatin stains deeply with basic dyes. *See* euchromatin and heterochromatin.

chromatography. A technique for separating and identifying the chemical compounds that make up mixtures of molecules.

chromatophore. In photosynthetic bacteria and some animals, a discrete vesicle delimited by a single membrane and containing photosynthetic pigments.

chromomere. Minute, spherical bodies in the chromosome thread.

chromonema. The thread of a chromosome sometimes visible as spiral thickening within the chromatids; composed of an inner core of DNA covered with protein.

chromophore.   The part of a phytochrome molecule that absorbs light; e. g. phycobilin pigments.

chromoplast.   A plastid containing pigments other than chlorophyll, usually yellow and orange carotenoid pigments; a colored plastid.

chromosome.   The self-duplicating structural units of the cell nucleus containing genes in a linear order; they are threads of chromatin which are normally dispersed between cell divisions, and become condensed into recognizable "chromosomes" only during mitosis and meiosis. Eukaryotic chromosomes contain a double helix of DNA, some RNA, and proteins (mostly histones). The genes of prokaryotic organisms are located on a double-stranded molecule of DNA. (*See* fig. S-1, S-11).

chytrid.   A phycomycete fungus of the order Chytridiales. Most are minute intracellular parasites or saprophytes.

cicatrix, *pl.* **cicatrices.**   A mark or scar left after abcission of a leaf or bract.

ciliate.   Fringed with conspicuous hairs along the margin.

ciliolate.   Diminutive of ciliate.

cilium, *pl.* **cilia.**   Minute short trichomes; with conspicuous marginal hairs. *See* ciliate.

cincinnus.   A tight, unilateral scorpioid cyme.

cinereous.   Ash-colored; light gray.

cion.   *See* scion.

circadian rhythms.   Regular rhythms of growth and activity, which occur at close to 24-hour intervals; they occur more or less independently of fluctuations in temperature and/or light. Circadian rhythms are evidence of the biological clock.

circinate, circinate vernation.   The coiled arrangement of leaves and leaflets in the bud, with the apex of the lamina in the center of the coil, as in ferns and cycads.

circular bordered pit.   A bordered pit with a circular aperture.

circumscissile.   Opening or dehiscing by a line around a fruit or anther, the valve usually coming off as a lid.

cirrus.   A curl, a tendril.

cismontane.   "This side" of the mountains, as opposed to the "other side"; e.g. on the

West coast of N. America, used in reference to the area west of the Cascade-Sierran Crest, as opposed to the deserts.

cisterna, *pl.* **cisternae.**   A portion of endoplasmic reticulum consisting of two parallel membranes and an enclosed space; or one of the flattened sacs that combine with others to form a dictyosome or Golgi body.

cistron.   The sequence of codons necessary to define the composition of a single protein. *Syn.* gene.

citric acid cycle.   *See* Krebs cycle.

cladistic.   Genetic relationships which reflect or imply recent common ancestry.

cladode, cladophyll.   A flattened foliaceous stem having the form and function of a leaf, but arising in the axil of a minute, bractlike, often caducous, true leaf. *Syn.* phylloclade. (*See* fig. C-7).

cladodromous.   A leaf with a single primary vein, and with secondary veins freely branched toward but not terminating at the margin.

cladoptosic.   Shedding of branches, stems and leaves simultaneously.

clamp connection.   In Basidiomycetes, a temporary hyphal outgrowth from the apex of a secondary dikaryotic mycelium. They are formed during cell division and ensure that two dissimilar nuclei occur in the hyphae of the basidiocarp.

clasping.   A leaf base which partly or wholly surrounds the stem.

Fig. C-7. Cladode: a. true leaf, b. flower bud, c. true leaf, d. cladode.

class.   A group of plants ranking above an order and below a division (phylum).

classification.   The systematic arrangement of plants into groups (taxa) based on characteristics common to the group, especially those resulting from a common descent. The taxa in common use, from the smallest to the largest and most inclusive are: species, genus, family, order, class, and division.

clavate, claviform.   Club-shaped; gradually thickened toward the apex from a slender base, like a baseball bat.

claviculate.   Furnished with tendrils or hooks.

claw.   1. The long narrow petiole-like base of the petals or sepals in some flowers. 2. The modified auricle of some grass leaves, such as wheat and barley.

cleavage embryogeny, cleavage polyembryony.   A process in some gymnosperms in which the original embryo divides to form four or more distinct embryos; usually only one survives.

cleft.   Cut into lobes separated by narrow or acute sinuses which extend one-quarter to one-half the distance to the midrib.

cleistogamous, cleistogamy.   Descriptive of a flower that does not open and is self-pollinated. Compare chasmogamous.

cleistogene.   A plant which produces cleistogamous flowers.

cleistothecium.   In Ascomycetes, a completely enclosed ascocarp containing asci and ascospores.

climax community.   The terminal stage of an ecological succession sequence, which remains relatively unchanged as long as climatic and physiographic factors remain stable.

climber.   A plant which grows upward using other plants or objects for support.

clinanthium, clinium.   The compound receptacle of the composite head.

cline.   A gradual and more or less continuous change in a phenotypic character over the geographical range of a species. Frequently individuals at the two extremes of these continuous populations are strikingly different, morphologically, physiologically, and perhaps cytogenetically. Such populations which differ slightly from one another are termed ecotypes.

clip.   The seizing mechanism in the flowers of asclepiads (milkweeds),

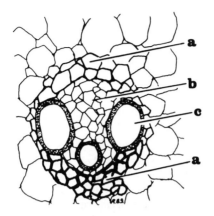

Fig. C-8. Closed bundle: a. bundle sheath, b. phloem, c. xylem vessel.

clone.   A group of genetically identical individuals resulting from asexual, vegetative multiplication, i.e. by mitosis; any plant propagated vegetatively and therefore considered a genetic duplicate of its parent.

closed bundle.   A vascular bundle in which a cambium does not develop, as in the bundles of monocot stems. (*See* fig. C-8).

closed venation.   Venation in a leaf characterized by anastomosing veins.

closing trap.   An active trap of a carnivorous plant which has two nearly identical halves which capture the prey.

club fungi.   The Basidiomycetes, which possess club-shaped basidia.

club moss.   A seedless vascular plant resembling true mosses, but in contrast to them, the sporophyte is dominant; e.g. *Lycopodium, Selaginella.*

cluster cups.   The aecia of rust fungi.

coal.   The carbonized vegetable material composed mainly of organic carbon, formed by the compression of plant remains from the carboniferous forests.

Coal age.   *See* Carboniferous period.

coalescence.   The union of like parts or organs.

cob.   The rachis of the pistillate corn (maize) spike.

coccoid.   A round or spherical cell type or morphological form.

coccus, *pl.* **cocci.   1.** A spherical bacterium. **2.** A berry; in particular, one of the parts of a lobed, sometimes leathery or even dry fruit with one-seeded cells.

cochlea.   A closely coiled legume.

cochleate.   Coiled like a snail shell.

codon.   A group of three adjacent nucleotides in a mRNA molecule that form the code for a single amino acid; sixty-four codons have been determined, of which three are used as stop signals to terminate the amino acid chain. *See* genetic code.

coefficient of coincidence.   An experimental value equal to the observed number of double crossovers divided by the expected number.

coelospermous.   Hollow-seeded.

coenobium.   1. A colony of unicellular organisms surrounded by a common membrane. **2.** A colony of green algal cells with the cell number fixed at time of origin.

coenocarpium.   Multiple fruit derived from ovaries, floral parts and receptacles of many coalesced flowers of an entire inflorescence; e. g. a fig.

coenocyst.   A multinucleate cyst or dormant spore.

coenocyte, coenocytic.   A multinucleate mass of protoplasm resulting from repeated nuclear divisions unaccompanied by plasma membrane and cell wall formation.

coenogamodeme.   A unit composed of all the hologamodemes which are considered capable of exchanging genes to some extent.

coenospecies.   A genecological term denoting ''the total sum of possible combinations in a genotype compound'' (Turesson).

coenozygote.   A multinucleate zygote.

coenzyme.   An organic molecule associated with and usually needed in an enzyme-catalyzed process; e.g. ATP, NAD and NADP are common coenzymes.

coetaneous.   With flowers and leaves developing at about the same time.

coevolution.   A dynamic process involving the reciprocal interactions between participating organisms profoundly affecting each others characteristics during the course of evolution.

coherent.   With like parts or organs joined, but only superficially and not actually fused.

cohesion.   1. The force which holds molecules of a substance together. 2. The union of like or similar parts.

cold frame.   An outdoor growing area that can be covered with glass or transparent plastic; used for starting seeds in early spring, or for hardening-off greenhouse-grown plants before transplanting them into the field or before sale.

cold-hardened.   A condition in which a plant is gradually exposed to, and has become increasingly resistant to cold conditions.

coleoptile.   A protective sheathlike structure enclosing the epicotyl in seeds of grasses. (*See* fig. R-1).

coleorhiza.   A protective sheathlike structure enclosing the radicle in grass embryos. (*See* fig. R-1).

coliform.   Designating or resembling a group of bacteria.

collar.   The transition zone between a primary stem and root. *Syn.* collet.

collateral bundle.   A vascular bundle having the phloem on one side of the xylem only.

collecting hairs.   Papillose or multicellular trichomes of stigmatic surfaces which collect pollen. Collecting hairs have various morphologies and are often secretory.

collenchyma.   A flexible supporting tissue composed of elongated living cells with unevenly thickened primary walls; often found in regions of primary growth in stems and some leaves.

collet.   *See* collar.

colleter.   A multicellular glandular appendage (often a trichome) which secretes a sticky substance.

colloid.   A permanent suspension of fine particles.

colony.   1. In algae, an aggregation of closely associated cells, the units of which function independently of each other but do not usually occur separately. 2. In bacteria, a mass of individuals, usually derived from cells of a single species. 3. An ecological term referring to a group of plants. 4. In fungi, refers to many hyphae projecting from a single point and forming a round or globose thallus.

colporate.   A pollen grain with compound apertures; e.g. a porate colpi.

colpate. A pollen grain bearing one or more colpi.

colpus, *pl.* **colpi.** An oblong-elliptic aperture or furrow in the exine of a pollen grain, running from one pole to another, with length/breadth ratio greater than two.

columella. A sterile structure within a sporangium or other fructification, as in mosses, liverworts and certain fungi; often an extension of the stalk. (*See* fig. C-9).

column. **1.** In orchids, the structure formed by the fusion of the stamens and carpels. *Syn.* gynostemium. **2.** In grasses, the basal, twisted portion of an awn. **3.** In mallows, the fused staminal tube. **4.** In carnivorous plants, the structure supporting the lid or hood of a pitcher plant leaf.

com-. A prefix meaning with or together.

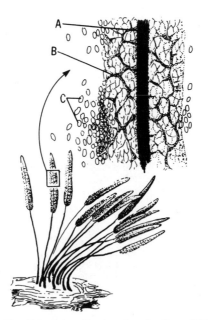

Fig. C-9. Columella: A. columella, B. capillitium.

coma.   **1.** The trichomes at the end of some seeds. **2.** The leafy crown or head as in many palm trees.

commensalism.   A symbiotic relationship between an organism (the commensal) and its host, in which the host neither benefits nor suffers from the association while the commensal benefits.

commissural bundle.   A small vascular bundle interjoining larger bundles in leaves of grasses.

commissure.   The place of joining or meeting.

community.   An assemblage of organisms living together and interacting with each other in a common environment; a group of populations of different species that live in the same ecosystem.

comose.   Bearing a tuft of trichomes, usually apically.

companion cell.   Parenchyma cells in the phloem of angiosperms; one or more are usually associated with a sieve tube member and originate from the same mother cell as the sieve tube member; they function in the translocation of sugars into, and out of, the sieve elements. (*See* fig. C-10).

companulate.   Bell-shaped.

comparium.   *See* syngamodeme.

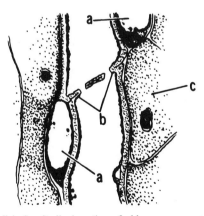

Fig. C-10. Companion cell in longitudinal section of phloem: a. companion cell, b. sieve plate, c. parenchyma cell.

compensation point.   The level of light intensity at which oxygen production during photosynthesis, is equal to oxygen consumption by respiration.

competition.   The simultaneous demand by two or more organisms on a limited environmental parameter, such as food, water, territory, a mate, etc.

complanate.   Flattened; compressed.

complementary cells, complementary tissue.   *See* filling tissue.

complete.   Having all the parts belonging to it, as a complete flower, with sepals, petals, stamens and carpels.

complicate.   Folded longitudinally.

composite.   **1.** A common name for plants in the sunflower family (Asteraceae). (*See* fig. C-2). **2.** Closely packed.

compound.   Having two or more similar parts in one organ.

compound inflorescence.   An inflorescence composed of secondary branches (inflorescences).

compound leaf.   A leaf of 2 or more leaflets; in some cases the lateral leaflets may have been lost and only the terminal leaflet remains. Compare simple leaf.

compressed.   Flattened laterally or from side to side.

compression.   A type of plant fossil formed when a plant part is deposited upon soft mud, then covered with a fine sediment. It is compressed from the weight of accumulated sediments; internal structures are not preserved but epidermal structures may be. (*See* fig. F-4).

compression wood.   The reaction wood of conifers which is found in leaning or crooked stems forming dense, lignified cells; it is formed on the lower sides of branches that are bent down (pushing them back up), or at the top of a branch bent up (pushing it down). Compare tension wood.

con-.   A prefix meaning with, together.

concentric.   *See* amphivasal.

conceptacle.   A depression in the thallus of certain brown algae in which gametes are produced.

conchoidal.   Having the form of half of a bivalve shell.

concolor. Of uniform color.

concrescent. Growing together of parts originally separate.

conduplicate. Longitudinally folded upward or downward along the central axis so that ventral and/or dorsal sides face each other.

cone. 1. A strobilus; a fruiting structure usually elongated, present in club mosses, conifers and cycads consisting of a group of sporophylls bearing sporangia. (*See* fig. L-6). **2.** An inflorescence or fruit with a central axis bearing overlapping scales.

confluent. Merging or blending of one part with another.

congeneric. Belonging to the same genus.

congested. Crowded. Contrast lax.

conglomerate. Densely clustered.

conical. Cone-shaped.

conidiophore. In fungi, a specialized hypha that bears conidia. (*See* fig. S-19).

conidium, *pl.* **conidia.** A spore formed asexually, usually at the tip or side of a hypha. (*See* fig. S-19).

conifer. A general term referring to the Coniferinae, one of four classes of gymnosperms; a cone-bearing tree or shrub, e.g. pine, fir, spruce, juniper, cypress.

conjugate nuclear division. The simultaneous mitotic division of the two nuclei of dikaryotic cells.

conjugation. 1. In eukaryotes, the union of two gametes. **2.** The recombination mechanism in bacteria that closely resembles sexual reproduction in other organisms.

conjugation tube. The tube connecting two conjugating cells in the green alga *Spirogyra;* through this tube, one gamete moves toward another gamete or both gametes move toward each other.

conk. The basidiocarp of a bracket fungus.

connate. Union or fusion of like parts or organs to one another with histological continuity.

connate-perfoliate. Opposite, sessile leaves, united at their bases, surrounding the stem.

connective.   An extension of the filament, connecting the two cells of an anther.

connivent.   Convergent without organic fusion.

conoidal.   Cone-shaped.

conopodium.   A conical floral receptacle.

conspecific.   Within or belonging to the same species.

constancy (of pollinators).   A term describing the behavior of an animal, usually an insect, which visits only a single plant species during a certain time period or on a given flight.

constricted.   Tightened or drawn together.

consumer.   In an ecosystem, a heterotrophic organism which feeds on other organisms.

context.   The fibrous tissue which makes up the body of the pileus of the club fungi (Basidiomycetes).

contiguous.   Touching without fusion; used irrespective of whether the parts are like or unlike.

continental drift.   The slow movement of continental land masses, composed of lighter rock materials, over the denser material beneath them.

continental island.   An island assumed to have been once connected to a neighboring continental land mass at some time in the geological past. Compare oceanic island.

continuous.   Symmetry of arrangement even, not broken.

continuous distribution.   Measurements taken from a population of values which form a continuous spectrum of values from one extreme to the other. Compare discontinuous variation.

continuum.   A term used in community analysis to designate the gradient containing ordered species or communities.

contorted.   Twisted around a central axis, bent or distorted.

contractile roots.   A specialized type of root, often found in bulbous plants, that undergoes contraction and thereby pulls the bulb or shoot parts deeper into the soil. They develop apparently in response to fluctuating temperatures near the soil surface when leaves become exposed to light; common among monocotyledons and also in certain herbaceous perennial dicotyledons (e.g. *Oxalis, Jepsonia,* etc.) and ferns (e.g. *Botrychium*).

contractile vacuole.   A vacuole located near the anterior end of many unicellular organisms which by expansion and contraction, functions in the elimination of excess fluids.

controlling gene.   A gene that controls the expression of a structural gene; there are two types: operator genes, which directly control structural genes, and regulator genes, which control operator genes.

convergent evolution.   The independent development of similar structures in unrelated or distantly related forms of life; often found in organisms living in similar environments, e.g. as seen in the desert plants in the Euphorbiaceae and Asclepidaceae in Africa and the Cactaceae in the New World. *Syn.* parallel evolution; convergence. (*See* fig. C-11).

convolute.   Rolled up longitudinally and usually twisted apically. Used in reference to leaf blades or floral envelopes in the bud when one edge is outside and the other inside.

coordinated growth.   Growth of cells in a manner that involves no wall separation, as opposed to intrusive growth. *Syn.* symplastic growth.

coprophilous.   Growing on, or living in dung.

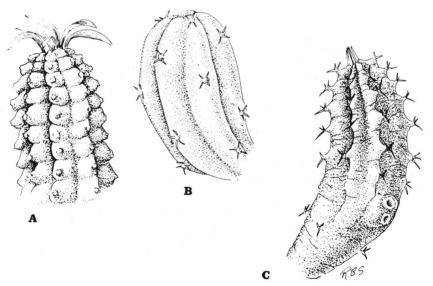

A

B

C

Fig. C-11. Convergent evolution in desert plants: A. Euphorbiaceae, B. Cactaceae, C. Asclepiadaceae.

coralloid.   Corallike.

coralloid roots.   In cycads, masses of apogeotropic roots which form near the surface of the soil; they are dichotomously branched and swollen from the presence of a blue-green (*Anabaena*) in the cortex. (*See* fig. S-18).

corbicula, *pl.* **corbiculae.**   A specialized structure found on the legs of certain bees used for the temporary storage and transportation of pollen.

cordate, cordiform.   **1.** Shaped like a stylized heart; ovate in general outline. **2.** In reference to leaves, having the notched end at the base, and the pointed end at the apex. *Opp.* obcordate.

core.   The central portion of a fleshy fruit, especially pome fruits.

coremium.   *See* synnema. (*See* fig. S-19).

coriaceous.   Thick, tough and leathery.

cork, phellem.   A secondary tissue produced by a cork cambium; polygonal cells, nonliving at maturity, with walls infiltrated with a waxy or fatty material (suberin) resistant to the passage of gases and water vapor. Replaces the epidermis in older stems and roots of many seed plants. (*See* fig. C-12).

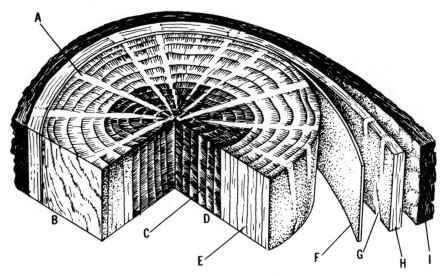

Fig. C-12. Cork: A. xylem ray, B. tangential surface, C. heartwood, D. radial surface, E. sapwood, F. vascular cambium, G. phloem ray, H. phloem (inner bark), I. cork (periderm).

cork cambium, phellogen.   A lateral zone of meristematic cells, outside the phloem, that produces cork cells external and phelloderm internally, forming the periderm. Found in the stems and roots of gymnosperms and many dicots.

corm.   A shortened (compressed) usually subterranean stem enclosed by dry, scalelike leaves; the bulk of the corm consists of storage tissue composed of starch-containing parenchyma cells. They produce two types of roots: fibrous and contractile roots.

cormel.   A small corm arising vegetatively from a mother corm.

corn.   **1.** In the United States used in reference to the fruits of *Zea mays.* **2.** In many other countries, it generally refers to the fruits of the principle cereal crop grown in a particular country.

corneous.   Horny; with the texture of horn.

corniculate.   Bearing or terminating in a small hornlike protuberance or process.

corolla.   A collective term referring to the petals of a flower; the inner circle or second whorl of floral envelopes; separate petals are said to be polypetalous; fused (connate) petals are said to be gamopetalous or sympetalous.

corolla tube.   A tubelike structure resulting from the fusion of the petals along their edges.

corolliform.   Appearing to be a corolla; as the calyx in the four o'clock family (Nyctaginaceae).

corona, coronate.   **1.** A crown; any appendage between the corolla and stamens which may be petaloid or staminal in origin. **2.** The crown of cells on the oogonium of members of the stoneworts (Charophyta).

coroniform.   Crown-shaped.

corpus.   A group of cells in the apical meristem which are located beneath the tunica and divide in various planes resulting in an increase in volume.

corrugate.   Irregularly folded or wrinkled.

cortex.   The primary tissue (or ground tissue) of the stem or root, located between the primary phloem (or endodermis, if present) and the epidermis; composed mostly of parenchyma cells, but many other cell types may also be present. (*See* figs. C-4, M-2, P-6).

cortication.   In some algae, the process in which lateral branches grow around the axis to form a cortexlike covering.

cortina. A term used to describe the partial veil of certain Basidiomycetes, where it is thin and diaphanous; as in *Cortinarius* spp.

corymb. A flat-topped or convex, indeterminate, racemose inflorescence, the lower or outer pedicels longer, their flowers opening first.

corymbose. Arranged in corymbs.

cosmopolitan. An organism which is essentially worldwide in distribution.

costa, *pl.* **costae.** **1.** In angiosperms, the midvein of a single leaf or the rachis of a pinnately compound leaf. **2.** In ferns, the midvein of a pinna or pinnule. **3.** In diatom frustules, a ridge or thickened rib. **4.** In mosses, the midrib of leaf or thallus. **5.** In pollen grains, a thickening of the nexine near an aperture.

costapalmate. A petiole which extends through the palmate leaf blade as a distinct midrib. Characteristic of some fan palms, as in the palmetto.

costate. Coarsely ribbed; with one or more longitudinal ribs or nerves.

cotyledon. An embryonic seed leaf or leaves which may function in food storage and may become photosynthetic when the seed germinates; there are two cotyledons in dicotyledons, in which they generally function in food storage; monocotyledons have a single cotyledon, in which food is generally absorbed from the endosperm and transferred to the embryo. In some plants, the cotyledon remains in the seed coats and in others, it emerges on germination. *See* hypogeal and epigeal.

cotyledonary node. The node on the embryonic axis where the cotyledon or cotyledons are attached.

cotyledonary trace. The leaf trace of a cotyledon, located within the hypocotyl.

cotyloid, cotyliform. Concave or cup-shaped.

coupling. Denoting a condition where linked, recessive alleles are found on one homologous chromosome while their dominant alternatives occur on the other homolog. *Opp.* repulsion.

covalent bond. A bond between atoms formed by the sharing of electrons.

covariance. The mean of the product of the deviation of two variates from their individual means. A statistical measure of the interrelation between variables.

cover. The amount of verticle projection of the crown or shoot area of a species in relation to the ground surface, expressed as a percentage; also, the projection of the basal area to the ground surface.

cover glass.   A very thin sheet of glass placed over material mounted on a slide to protect the material and aid in its microscopic examination.

crampon.   A hook or adventitious root which acts as a support.

craspedodromous venation.   Leaf venation characterized by a single primary vein with the lateral veins terminating at the margin.

crassinucellate ovule.   One of two general types of nucellar organization found in angiosperms in which the megasporocyte arises from a sporogenous cell deeply embedded within the nucellus. Compare tenuinucellate.

crassula, *pl.* **crassulae.**   Thickenings of intercellular material and primary wall along the upper and lower margins of a pit-pair, as found in the tracheids of gymnosperms. *Syn.* bars of Sanio.

crateriform.   Saucer- or cup-shaped; shallow.

creeper, creeping.   A nontechnical term for a trailing plant or shoot, which roots adventitiously along most of its length or at the nodes.

cremocarp.   A dry, dehiscent, 2-seeded fruit of the umbel family, each half a mericarp borne on a hairlike carpophore; a schizocarp.

crenate.   Shallowly ascending round-toothed, or teeth obtuse; teeth cut less than one-eighth way to midrib or midvein.

crenulate.   Minutely or finely crenate. Teeth sinuses extend one-sixteenth of the distance to midrib of leaf.

crest.   **1.** A ridge or elevation on a structure. **2.** In some milkweed flowers a hornlike projection from a segment (hood) of the corona.

crested, cristate.   **1.** Having a crest, elevated appendage, or ridge on the summit of an organ. **2.** An abnormal plant growth resulting from the extension of a single apical meristem into an apical ridge meristem. (*See* fig. C-13).

Cretaceous period.   The third and final period of the Mesozoic era characterized by great swamps, formation of the Rocky Mountains, the rise of flowering plants, modern insects, archaic mammals, and birds, and the extinction of dinosaurs, pterodactyls, and toothed birds.

crinite.   Bearded with long, weak hairs.

crispate, crisped.   Irregularly curled leaf margin or trichome.

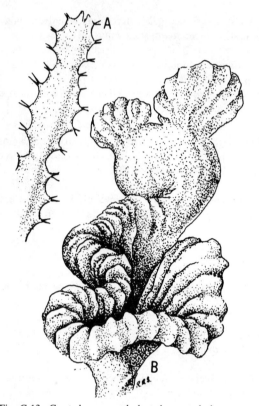

Fig. C-13. Crested: a. normal plant, b. crested plant.

crista, *pl.* **cristae.   1.** A crest or ridge. **2.** An invagination (infolding) of the inner membrane in mitochondria.

critical day length, critical light period, critical photoperiod.   The photoperiod required to induce flowering.

crop rotation.   The practice of growing different crops in regular succession as a method of controlling the spread of insects and diseases, to increase soil fertility, and to decrease erosion.

cross.   To interbreed; to mate two individuals of different breeds, races or species; the product of such a mating.

cross-fertilization.   The union of gametes of two different individuals.

cross-field.   A term of convenience for the rectangle formed by the walls of a ray cell and an axial tracheid, as seen in the radial section of conifer wood.

crossing-over. The exchange of corresponding segments between chromatids of homologous chromosomes during prophase I of meiosis, resulting in the recombination of linked genes. Chiasmata are the visible evidence of crossing-over.

cross-pollination. The transfer of pollen from the anther of one plant to the stigma of a flower of another plant.

cross section. A section cut perpendicular to the longitudinal axis of a structure. *Syn.* transverse section.

crown. **1.** The persistent base of a herbaceous perennial. **2.** The top of a tree; a corona. **3.** The junction between a stem and root in a seed plant. **4.** A part of a rhizome with a large bud, suitable for use in vegetative propagation. **5.** The circle of appendages on the throat of a corolla. **6.** A hard ring or zone at the top of the lemma in certain grasses.

crownshaft. In some species of palms, a glossy, green pillarlike extension of the trunk, formed by the long, broad, overlapping petiole bases.

crozier. Any plant structure with a curled end; as the unfolding pinnae of cycad and fern leaves. (*See* fig. C-14).

crozier formation. In the ascogenous hyphae of the Ascomycetes (sac fungi), formation of a hook in which conjugate nuclear division occurs, followed by cytokinesis.

cruciate, cruciform. Cross-shaped; e.g. four separate petals in the form of a cross, typical of the flowers of the mustard family (Brassicaceae).

cruciate basidium. In the heterobasidiomycetes, a basidium whose base contains longitudinal septations that are at approximate right angles to each other.

Fig. C-14. Fern crozier.

crucifer. A plant with four petals and tetradynamous stamens; a member of the mustard family (Brassicaceae).

cruciform. Intranuclear division in which the chromosomes are arranged in a ring around a dumbbell-shaped nucleolus as they divide. *See* cruciate.

crustaceous. Hard, thin and brittle.

crustose. A type of lichen that forms a thin, flat crust strongly attached to the surface on which it is growing. (*See* fig. L-4).

crymophilous, crymophytes. Plants which live in tundra or in polar regions.

cryoflora. Plants which live on ice and snow, especially certain algae.

cryotropism. Movement induced by cold or frost.

cryptanthous. With hidden flowers; cleistogamous; the stamens remaining enclosed in the flower.

cryptoblast. A sterile invagination, similar in structure to a conceptacle, found on the surface of brown algae (Phaeophyta).

cryptocotylar. Having the cotyledons remaining inside the seed coat; cotyledons usually remaining below ground. Contrast phanerocotylar.

cryptogam. A plant reproducing by spores instead of by seeds, as ferns, mosses, algae. Compare phanerogam.

cryptomonads. Microscopic, unicellular golden brown algae, similar to dinoflagellates.

cryptophyte. A plant which reproduces by underground or underwater structures, as by bulbs, corms or rhizomes.

crystalliferous cell. A specialized type of plant cell containing one or more crystals.

cucullate. Hooded or hood-shaped.

cucullus. A hoodlike process on some seeds; e.g. a caruncle.

culm. The stem of a grass or sedge.

cultigen. Horticultural plants known to exist only in cultivation. They presumably originated under domestication, e.g. maize and cabbage.

cultivar. A contraction of "cultivated variety". It refers to a plant type within a particular cultivated species that is distinguished by one or more characters; horticulturally, such plants are of considerable economic importance.

cuneate, cuneiform. Wedge-shaped; triangular, with the narrow part at the point of attachment.

cup, cupulate, cupuliform. A shallow, open structure, as the involucre of an acorn.

cup fungi. A type of ascomycete characterized by conspicuous cup- or saucer-shaped fruiting bodies called apothecia; most are saprophytic, but some are parasitic, including many important leaf pathogens, and others are symbionts of algae in lichens. *Syn.* Discomycetes.

cupule. A cuplike structure at the base of some fruits, formed by the fusion of involucral bracts at their bases; e.g. the acorn of oaks and the fruits of some palms.

cusp. An abrupt, sharp, often rigid point.

cuspidate. With an apex somewhat abruptly and sharply constricted into an elongated, sharp-pointed tip or cusp.

cuticle. A noncellular layer of waxy or fatty materials on the outer walls of epidermal cells. *See* cutin. (*See* fig. M-4).

cuticularization. The process of formation of cuticle.

cuticular respiration. The loss of small amounts of water vapor from leaves by direct evaporation from the epidermal cells through the cuticle.

cutin. A complex waxy substance that impregnates the walls of epidermal plant cells, and also occurs as a separate layer, the cuticle, on the outer surface of the epidermis. It renders the epidermis more or less impervious to water.

cutting. A vegetative portion removed (cut) from a plant for the purpose of propagation.

cyanophycean starch. A unique starch produced by the blue-greens.

cyathiform. Cup-shaped.

cyathium. A cuplike involucre enclosing flowers; an inflorescence characteristic of *Euphorbia* (Euphorbiaceae), in which the cyathium includes one carpellate and numerous staminate flowers; there are glands around the edges of the cyathium which frequently have petallike appendages (petaloid glands). (*See* fig. C-15).

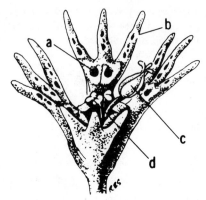

Fig. C-15. Cyathium: a. gland, b. petaloid appendage, C. pistillate flower, d. staminate flower.

cybernetics.   The science of controls, both animate and inanimate, which depend upon positive and negative feedback; cybernetics has important applications in applied ecology.

cycad.   A general term for the Cycadinae, one of four classes of gymnosperms. They are vascular plants with stout, mostly slow growing stems, often large compound leaves, and in general resemble palms. They are members of ancient gymnosperm groups and are strictly dioecious, producing male and female cones (except *Cycas*) on separate plants. They are unique with *Ginkgo* among seed plants in having motile sperms.

cyclic.   Whorled, the opposite of spiralled.

cyclochore.   A plant, whose seeds are disseminated by the wind tumbling the plant, as in Russian thistle (*Salsola*).

cycloid model.   A model of the eukaryotic chromosome having many loops of somewhat circular form attached to a central strand of DNA.

cyclosis.   The streaming of cytoplasm within a cell. *Syn.* cytoplasmic streaming.

cymba.   A woody boatlike spathe or spathe valve that encloses the inflorescence of many palms.

cymbiform.   Boat-shaped.

cyme.   A type of inflorescence consisting of a broad, more or less flat-topped, determinate flower cluster, with the central flowers opening first.

cymule.   Diminutive of cyme, usually few-flowered.

cynarrhodium.   A type of fruit characteristic of roses; also called a rose hip.

cypsela.   An achene derived from a one-loculed, inferior ovary; as in the indehiscent fruits of the sunflower family (Asteraceae).

cyst.   A resistant cell surrounded by an especially thick cell wall.

cystidium, *pl.* **cystidia.**   Specialized cells which occur with the basidia on the hymenium of certain Basidiomycetes; their function is unknown.

cystocarp.   In red algae, a spore case containing diploid spores (carpospores).

cystoliths.   Intercellular concretions which develop within the epidermal cells of certain plants; usually composed of calcium carbonate. (*See* fig. C-16).

-cyte, cyto-.   A suffix or prefix meaning "pertaining to a cell".

cytobiota.   A cellular entity; for example, the cell of a prokaryotic or eukaryotic organism.

cytochimera.   A type of chimera in which the cells in adjacent areas of a plant part have different numbers of chromosomes; e.g. where some cells are diploid and other polyploid.

cytochrome.   A series of pigmented, iron containing, protein molecules in mitochondria and plastids, involved in phosphorylation and photophosphorylation.

cytodeme.   A group of individuals of a taxon differing from others cytologically, usually in chromosome number.

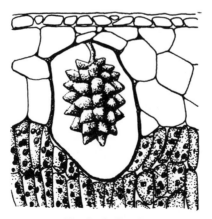

Fig. C-16. Cystolith.

cytogenetics.   The combined study of chromosome cytology with genetical methods and results of breeding analysis.

cytogeography.   The study of the relationship between the chromosomes (karyology) of a plant taxon and its geographical distribution.

cytohistological zonation.   The presence of areas in an apical meristem with distinctive cytological characteristics.

cytokinesis.   Division of the cytoplasm that often occurs during division of the nucleus (karyokinesis); if simultaneous with karyokinesis, cell division results. In certain algae and fungi, cytokinesis occurs only in reproductive parts.

cytokinin.   A class of hormones important in many growth responses of plants; e.g. promotes and controls rates of cell division (cytokinesis), promotes cell enlargement, bud and root formation, breaks seed dormancy, effects flowering photoperiod and induces parthenocarpy.

cytology.   The study of cells including both morphological and physiological investigations.

cytomixis.   The apparent migration of chromatin from one microspore (pollen) mother cell into another; also, the fusion of entire, adjacent microspore mother cells which may develop into polyploid gametes.

cytoplasm.   The protoplasm in a cell exclusive of that within the nucleus (nucleoplasm); includes the various organelles and membranes present in a cell. *Syn.* hyaloplasm.

cytoplasmic inheritance.   A non-Mendelian inheritance involving self-replicating, cytoplasmic organelles such as mitochondria, chloroplasts, viruses, etc.

cytoplasmic streaming.   The movement or flow of cytoplasm within a cell, its function possibly being to increase the rate at which dissolved substances move within the protoplasm. *Syn.* cyclosis, protoplasmic streaming.

cytotaxonomy.   The use of chromosomal relationships as an aid to determining taxonomic and evolutionary relationships.

cytotype.   A plant or group of plants distinguished from others of the species by some cytological feature.

D

damping off. A fungal disease of seedlings which causes them to rot and shrivel at soil level, or to die before they emerge from the soil.

dasycarpous. Thick-fruited.

dasyphyllous. Having thick leaves.

day neutral plant. A plant that flowers independently of day length (i.e. not stimulated by long or short days), with flowering controlled by other factors.

D.B.H. *See* diameter breast height.

de-. Prefix meaning down, away from or off.

deca-. A prefix meaning ten.

deci-. A prefix meaning one-tenth or ten.

deciduous. **1.** The falling of parts at the end of a growing period, as with leaves that are shed in autumn, or of floral parts that fall after anthesis. **2.** Referring to those plants (trees or shrubs) that drop their leaves at the end of each growing season. Compare evergreen.

declinate. Bent or directed downward or forward.

decomposers. A group of organisms, chiefly saprophytic fungi and bacteria, that break down organic material into smaller molecules.

decompound. A general term for leaves that are more than once divided or compound.

decortication. Removal of the bark or outer covering of a plant or plant product.

decumbent. A growing habit in which a portion of the stems or shoots lie close to the ground without rooting adventitiously. The upper parts of the stem are erect or ascending.

decurrent. An extension of tissue occurring down the stem below the point of insertion of a leaf petiole or ligule, forming a wing or ridge.

decussate. Opposite leaves alternating at right angles with those above and below. (*See* fig. D-1).

105

Fig. D-1. Decussate.

dedifferentiation.   The process in which mature cells become less mature and resume meristematic activity, as adventitious buds which arise from mature tissues.

deficiency.   *See* deletion.

definite.   Not exceeding twenty, with reference to stamens; as opposed to numerous.

deflexed.   Reflexed; bent or turned abruptly downward.

defoliation.   Loss of leaves, resulting from natural shedding or from the effects of insects.

deforestation.   Removal of forests from an area.

degrees of freedom.   The number of quantitative data points that vary freely and independently.

dehardening.   A decrease in the cold resistance of a plant, resulting from continuous exposure of cold hardened plant tissues to warm temperatures.

dehisce.   To split open when ripe, usually along definite lines or sutures to release seeds or spores.

dehiscence.   The method or process of opening of a fruit (seed pod), anther or sporangia.

dehiscent.   A structure (e.g. fruit, sporangium or anther) that splits open at maturity to release its contents.

deletion.   The loss of absence of a segment of a chromosome with its genes. *Syn.* deficiency.

deliquescent. **1.** A mode of branching in trees in which the trunk divides into many branches leaving no central axis, e.g. elms. **2.** Softening, liquefying or wasting away of tissues, as in the gills of mushrooms.

deltoid. Shaped like an equilateral triangle, but often with the sides a little curved toward the apex.

deme. A population composed of organisms that are likely to interbreed.

-deme. A biologically and taxonomically neutral suffix, denoting a group of individuals of a specified taxon. The prefix used with deme indicates a specific type of relationship between the individuals, e.g. topodeme, gamodeme, cytodeme.

denaturation. A change in the natural molecular configuration of an organic substance as a result of heat or chemical treatment or extreme changes in pH; usually results in a loss of biological activity.

dendritic. Branching like a tree; as the hairs of some members of the mustard family (Brassicaceae).

dendrochore. An area with trees.

dendrogram. A pictorial method of expressing the taxonomic relationships of one taxon to another on the basis of overall resemblance, but without any phylogenetic implications.

dendrograph. An instrument which is used to measure the periodic variations in the diameter growth of trees.

dendroid. Treelike; resembling the shape of a tree.

dendroid colony. A branching colony of cells.

dendroid venation. A type of leaf venation in which the minor veins do not form closed meshes around small areas of mesophyll.

dendrology. The study of trees and shrubs.

dendrophilous, dendrophytes. Plants which live on trees or in forests.

denitrification. The process by which nitrogen is released from the soil by the action of denitrifying bacteria.

denitrifying bacteria. Bacteria capable of converting nitrates into nitrites, oxides of nitrogen, or free nitrogen. (*See* fig. N-2).

de novo. 1. Arising from an unknown source. **2.** Synthesized from very simple precursors.

density. The actual number of individuals in a given unit area.

density-dependent. Refers to any numerical property which changes with a change in the density of organisms or of species; e.g. birth rate or death rate.

density-independent. Refers to any numerical property which does not change and is independent of changes in the density of organisms or of species.

dentate. A leaf margin with sharp teeth or indentations pointing outwards at right angles to the midrib.

denticidal capsule. One that dehisces apically, leaving a ring of teeth.

denticulate. Minutely or finely dentate; the diminutive of dentate.

denuded. Naked or nearly so.

deoxyribonucleic acid (DNA). The carrier of genetic information (genes) in cells, composed of chains of phosphate, sugar molecules (deoxyribose), and purine and pyrimidine bases; the DNA molecule is capable of self-replication.

deoxyribose. A sugar characteristic of DNA.

depauperate. Dwarfed, stunted or poorly developed.

depilation. The natural loss of pubescence by plants as they mature.

deplanate. Flattened or expanded.

depolymerization. The breakdown of organic compounds into two or more, less complex molecules.

depressed. Pressed downward close to the axis with an angle of divergence of 166 to 180°; flattened endwise or from above.

depurlation. The loss of bud scales as the result of leafing out.

derivative. A cell produced by division of meristematic cells.

derived. An organism descended or modified from an older precursor.

derma, *pl.* **dermata.** The outer surface of an organ, e.g. the bark, rind or skin.

dermal tissue system. The epidermis or periderm; the outer covering tissue of a plant.

dermatogen. *See* protoderm.

dermatomycosis. A fungal infection of the skin of man or other animals.

dermatophytes. Fungi which cause diseases of the skin, hair and nails.

desalination. The process of removing salts from marine or brackish water, or from soil.

descending. Directed or tending to grow downward, with an angle of divergence of 136° to 165°, as in the branches of certain trees.

desert. A region characterized by scant rainfall, or by lack of available water.

desiccant, desiccation. 1. Any chemical used to dry plant materials; a drying agent. 2. To dry thoroughly by removing moisture.

desmid. Unicellular green algae (Chlorophyta), common in fresh water. Most desmids are composed of two semicells connected by a region called an isthmus.

desmogen. Meristematic tissue that differentiates into a vascular bundle.

desmotubule. A tubule which connects two endoplasmic reticulum cisternae, located at opposite ends of a plasmodesma.

desynapsis. The early separation of previously paired chromosomes before completion of meiotic prophase.

detached meristem. An axillary meristem which gives rise to an axillary bud and appears detached from the apical meristem because of intervening vacuolated cells.

determinate growth. Growth of limited duration, characteristic of floral meristems and leaves. Contrast indeterminate growth.

determinate inflorescence. An inflorescence in which the terminal or central flower develops first, thereby arresting further elongation of the inflorescence axis. Contrast indeterminate inflorescence.

detritus. Any particulate accumulation of disintegrated animal, mineral or vegetable debris.

deuteromycete. Any fungus of the Deuteromycetes which comprises the Fungi Imperfecti, so called because the sexual stages of their life cycle are lacking or unknown.

development.   The process of change in form and complexity of a whole plant or organism, or part of a plant or organism, from a juvenile to a mature state. *See* differentiation.

devernalization.   The process of reversing vernalization by application of a heat treatment.

deviation.   Departure of an observation from its expected value.

dextrose.   **1.** Rising helically or turning from right to left, characteristic of twining stems. **2.** Glucose.

d.f.   An abbreviation for degrees of freedom.

di-.   A prefix meaning two.

diacytic stoma.   A stoma which is enclosed by two subsidiary cells whose common wall is at right angles to the axis of the guard cells. (*See* fig. S-16).

diadelphous.   Stamens united (connate) by their filaments into two bundles or clusters; e.g. as in many legumes with nine stamens in one bundle and a single separate stamen.

diadromous venation.   Venation shaped like a fan, as in the leaf of *Ginkgo biloba* or in the pinnae of maidenhair ferns (*Adiantum*).

diageotropic.   Transversely geotropic, said of leaves or roots which grow horizontally.

diakinesis.   The last stage of prophase I of meiosis, characterized by the maximal contraction of the diplonema chromosomes. By end of this phase, the nucleoli and nuclear membrane disappear and the spindle is found.

dialycarpic.   Having a fruit composed of distinct carpels.

dialypetalous.   Polypetalous, the corolla being composed of separate and distinct petals.

dialysis.   The separation of different sizes of molecules in a mixture by their differential diffusibility through a porous membrane.

diameter breast height (DBH).   The diameter 4.5 feet (1.37 meters) above ground level.

diandrous.   With two stamens per flower.

diaphanous.   Transparent, translucent.

diaphragm, diaphragmed pith.   A dividing membrane or partition; a solid core of pith cells transversed by distinct layers (diaphragms) of firm-walled cells, forming small compartments in the pith.

diarch. Having two protoxylem poles or strands, as in the primary xylem of roots.

diaspore. *See* disseminule or propagule.

diatom. A unicellular, filamentous, or colonial microscopic alga with a cell wall composed of silica and divided into two valves which fit together, one overlapping the other. Centric diatoms possess a radially symmetrical body. Pennate diatoms possess bilateral symmetry. (*See* fig. D-2).

diatomaceous earth. Earth composed of the accumulated deposits of the siliceous cell walls of diatoms.

dicarpellary. Having two carpels.

dicaryon, dicaryotic. *See* dikaryon.

dicentric. A chromosome or chromatid with two centromeres.

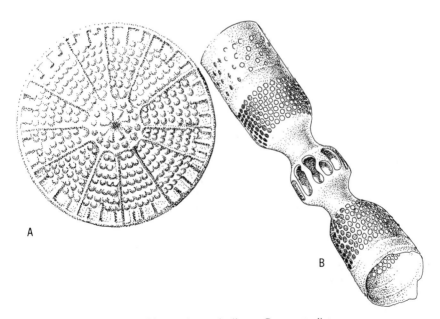

A

B

Fig. D-2. Diatom: A. centric diatom, B. pennate diatom.

dichasium.   A determinate type of cymose inflorescence having a central older flower which develops first,   and a pair of opposite lateral branches bearing younger flowers.

dichlamydeous.   Having a perianth composed of a distinct calyx and corolla.

dichogamous, dichogamy.   In a perfect flower, maturation of stamens and stigma at different times, thus preventing self-pollination.

dichotomous, dichotomy.   Branching by repeated forking in pairs; the division or forking of an axis into two, more or less equal, branches; e.g. dichotomous venation, and dichotomous branching.

dichotomous key.   In taxonomy, a key in which each division is divided into two subdivisions.

diclesium.   An achene or nut enclosed within a free but persistent calyx.

diclinous.   Having staminate and pistillate flowers either on the same plant (monoecious), or on different plants (dioecious).

dicot.   An abbreviated term for dicotyledon.

dicotyledon.   A member of the Dicotyledoneae; one of the two taxa (class or subclass), of angiosperms usually characterized by the following: two seed leaves (cotyledons), flower parts in fours or fives (or multiples of these), leaves with net venation, and root systems with tap roots. Compare monocotyledon.

dicotyledonous.   With two cotyledons.

dictyosome.   *See* Golgi body.

dictyospore.   A spore with both vertical and horizontal septa.

dictyostele.   Stele with a cylindrical arrangement of xylem and phloem together, in separate vascular bundles; each individual bundle termed a meristele. This arrangement is formed by large overlapping leaf gaps which dissect the vascular system into strands; common in ferns and in some juvenile plants, e.g. *Ginkgo biloba*.

dicyclic.   A series of organs arranged in two whorls, as a perianth of calyx and corolla.

didymous.   In two equal pairs, as with fruits or stamens.

didynamous.   With four stamens in two pairs of two different (unequal) lengths, as in most members of the mint family (Lamiaceae).

dieback.   A progressive death of plant shoots beginning at the tip.

differentially permeable membrane. A membrane through which substances diffuse at different rates, by allowing some to pass through more readily than others. A membrane which permits the free passage of water molecules but restricts the passage of dissolved solutes, as do all living membranes. This term is preferable to semipermeable membrane.

differentiation. The physiological and morphological changes that occur in a cell, tissue or organ during development from a juvenile to a mature state. *See* development.

diffuse. Loosely branching or spreading; widely spread, as in diffuse roots.

diffuse apotracheal parenchyma. Axial parenchyma occurring as single cells in wood, or irregularly distributed in strands among fibers, as seen in cross section.

diffuse centrome. Applied to chromosomes which lack a single, localized centromere.

diffuse growth. Cell division and elongation occurring throughout the plant.

diffuse porous wood. A wood in which the pores or vessels are fairly uniformly distributed throughout the growth layers or in which the size of pores changes only slightly from early wood to late wood. (*See* fig. D-3).

diffuse root system. *See* fibrous root system.

diffusion. The movement of suspended or dissolved particles from a more concentrated to a less concentrated region of the diffusing material as a result of the random movement of individual molecules; the process of random movement tends to distribute molecules uniformly throughout a medium. Compare active absorption.

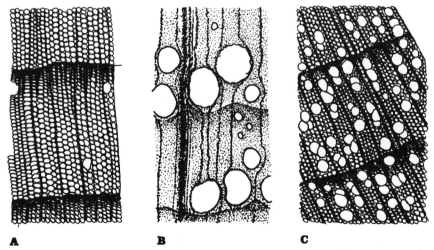

A   **B**   **C**

Fig. D-3. Diffuse porous: A. nonporous wood, B. ring porous wood, C. diffuse porous wood.

diffusion pressure.   The pressure exerted by a diffusing substance, proportional to its concentration.

diffusion pressure deficit (DPD).   The amount by which the diffusion pressure of the water in a solution or in soil is less than that of pure water at the same temperature and atmospheric pressure.

digestion.   The conversion of complex, usually insoluble foods into simple, usually soluble and diffusible forms, by means of enzymatic action requiring energy.

digitate.   Fingerlike; shaped like an open hand; palmate.

dihybrid.   The offspring of a cross between parents differing in two pairs of heterozygous genes.

dikaryon, dikaryotic.   A condition in the mycelium of Basidiomycetes and in ascogynous hyphae of Ascomycetes, in which each hyphal segment contains two unfused nuclei, the result of fusion of hyphae from two different mating types. Also spelled dicaryon, dicaryotic.

dilated.   Flattened and broadened, as an expanded filament.

dimidiate.   Divided in two, but with one part so much smaller as to appear wanting; unequally divided.

dimorphic, dimorphism.   Occurring in two distinct forms, sizes or shapes within the same species; e.g. having two forms of leaves, juvenile and adult, or having two kinds of pollen produced by the same plant.

dinoflagellate.   Members of the Pyrrophyta, mostly unicellular motile biflagellates, which occur in both fresh and marine waters; sometimes forming red tides that kill fish. (*See* fig. D-4).

dioecious, dioicous.   1. Having staminate and pistillate flowers on different plants of the same species; thus dioecious plants have imperfect and unisexual flowers. A term properly applied to a taxon and not to flowers. 2. Having male and female organs on different plants.

dioicous.   Same as dioecious.

diplanetism.   The occurrence of two motile periods, interrupted by a resting stage in a life cycle, as in the water molds.

diplecolobal.   Having incumbent cotyledons folded two or more times.

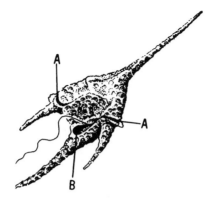

Fig. D-4. Dinoflagellate: A. girdle, B. sulcus.

diplobiontic.   Having two free-living thalli in the life cycle, one haploid, the other diploid.

diploid.   A nucleus, organism or generation which has two sets of chromosomes and thus has the 2n chromosome number, characteristic of the sporophyte generation. Compare haploid.

diploidization.   **1.** In fungi, the fusion of two haploid nuclei to form a diploid nucleus. **2.** An established polyploid which has normal meiosis and otherwise acts as a normal diploid.

diplonema.   *See* diplotene.

diplophase.   *See* diplotene.

diplospory.   The condition in some higher plants in which a diploid embryo sac is formed directly from a megaspore mother cell; an embryo is then formed without fertilization; agamospermy.

diplostemonous.   Having the stamens in two whorls, the outer whorl alternate with petals, or opposite the sepals, and the inner whorl opposite the petals.

diplotegium.   A pyxis derived from an inferior ovary.

diplotene.   The fourth stage of prophase I of meiosis. During this stage, the homologs begin to separate, but remain attached at chiasmata regions. *Syn.* diplonema and diplophase.

dipole.   A molecule which has opposite charges at opposite poles.

dipterous.   Having two wings.

directional selection.   Selection occurring when the environment is changing in a systematic fashion, leading to a regular change in a particular direction.

disarticulate, disarticulating.   To separate at a preexisting point; e.g. disarticulation above or below the glumes in grasses.

disc, disk.   **1.** A somewhat fleshy structure developed from the receptacle at the base of the ovary, or from coalesced nectaries or stamens around the pistil. **2.** The central part of the head of composites, as opposed to the rays. **3.** The flattened tip of a tendril.

disc flower.   *See* disk flower.

disciform.   Shaped like a disk, but depressed or shallow.

discoid.   **1.** In the Asteraceae, a head with disk flowers, and lacking ray flowers. **2.** Resembling a disk, orbicular with convex faces.

discontinuous.   Not continuous; having gaps or interruptions.

discontinuous distribution.   **1.** The occurrence of related organisms (e.g. members of the same species) in widely separated geographical areas. **2.** Data that are recorded as whole numbers, and thus do not result in a continuous series of values. Compare continuous distribution.

discrete.   Separate, not coalesced.

disease.   A condition in which a plant is adversely affected in its growth or functioning by another organism (a disease organism) such as fungi, bacteria, virus or nematodes.

disharmonic biota.   Refers to a biota which contains only a small fraction of the basic adaptive types available in the surrounding source regions, e.g. dispersal types. Compare harmonic biota.

disjunct distributions.   The occurrence of related or identical organisms in widely separated geographical areas.

disjunction.   The separation of homologous chromosomes during anaphase II of meiosis.

disjunctor.   A cell or projection which connects spores in a chain.

disk.   *See* disc.

disk flower.   The tubular, actinomorphic flowers in the center of the inflorescence (head) of plants in the sunflower family (Asteraceae). Contrast ray flower. (*See* fig. C-2).

dispersal.   To scatter or disperse, as with seeds, spores, pollen, etc. *Syn.* dissemination.

dispersal unit.   Any detached part of a plant involved in dispersal. May be a spore, seed, fruit or portion of a vegetative plant body. *Syn.* disseminule.

displacement.   *See* character displacement.

disruptive selection.   Selection which breaks up a previously homogenous population (gamodeme) into several distinct groups, each subject to different selection pressures.

dissected.   Deeply divided or irregularly cut into many segments.

dissemination.   *See* dispersal.

disseminule.   **1.** A plant part that gives rise to a new plant. **2.** Agents of dispersal, characteristic of a species, such as winged or barbed fruits. *Syn.* diaspore, dispersal unit.

distal.   Opposite from the point of origin or attachment; toward the apex.

distal face.   That part of a spore surface which is directed outwards in its tetrad.

distichous.   Two-ranked, with leaves or flowers in two opposite rows in the same plane.

distinct.   Separate; not united with similar organs.

distribution.   The range or geographical area inhabited by a species.

distromatic.   Having a thallus which is two cells thick.

distylous, distyly.   Referring to the flowers of a species which possess one of two style types: a long style (''pin'' flowers), and a short style (''thrum'' flowers). These function as a mechanism to promote outcrossing. (*See* fig. D-5).

diurnal.   Opening during the day; occurring during daytime.

divaricate.   Widely spreading.

divergent.   Spreading broadly with an angle of divergence of 15° or less; spreading less than divaricate.

diversity.   **1.** A measure of the total number of species present. **2.** A measure of the number of species present and their relative abundance.

divided.   Separated by deep cuts or incisions to, or near to the base.

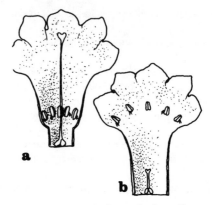

Fig. D-5. Distylous: a. pin flower, b. thrum flower.

division.   The largest category in the classification of plants, equivalent in rank to phylum in the animal kingdon.

divisural line.   The medial line along which the peristome teeth split.

DNA.   *See* deoxyribonucleic acid.

DNA hybridization.   A technique that compares the nucleotide sequence in DNA from different organisms; useful in establishing the amount of repetitive DNA in the genome of an organism.

dodecamerous.   In twelve parts; as in possessing twelve pistils (dodecagynous) or twelve stamens (dodecandrous).

dolabriform, dolabrate.   Axe-shaped or hatchet-shaped. *Syn.* asciiform.

doliform.   Barrel-shaped.

dolipore septum.   A septum which flares out in the middle portion of a hypha, forming a barrel-shaped structure with open ends.

domestication.   To adapt wild plants for cultivation by man.

dominant.   **1.** A gene that expresses its full phenotypic effect by masking the effect of its allelic partner. **2.** An ecological term for a species which due to its size and/or numbers, determines the character of a community.

donor parent.   The parent from which one or more genes are transferred to the recurrent parent in backcross breeding.

door.   In the bladderworts (*Utricularia*), a veil of tissue which closes the trap opening.

dormancy, dormant.   A period of inactivity in spores, seeds, bulbs, buds and other plant organs in which active growth has temporarily ceased.

dorsal.   1. The upper surface of a plant body having dorsiventral surfaces, as in liverworts. **2.** The lower, or under surface of a leaf. *Syn.* abaxial. *Opp.* ventral.

dorsifixed.   Attached medially to the back or dorsal surface, as with anthers of *Lilium*.

dorsilaminar.   On the dorsal side of a leaf blade.

dorsiventral, dorsiventral symmetry.   Flattened and having distinct dorsal and ventral surfaces, as a leaf. *Syn.* dorsoventral.

double.   Said of flowers that have more than the usual or normal number of floral envelopes, particularly petals.

double cross.   A cross between two F_1 hybrids.

double fertilization.   In angiosperms, a process in which each of the two sperm nuclei formed in the pollen tube fuses with a nucleus in the embryo sac. One sperm fuses with the egg nucleus to form the diploid (2n) zygote and the other sperm fuses with one or more polar nuclei to form the endosperm.

double-serrate.   With coarse serrations bearing minute teeth on their margins.

down, downy.   Covered with short, weak hairs or pubescence, as on the pappus of certain sunflowers, or the fruit and leaves of certain plants.

drepanium.   A flattened and coiled or curved monochasial inflorescence; a sickle-shaped cyme.

drift.   Changes in gene and genotypic frequencies in small populations due to random processes. *See* genetic drift.

drupaceous.   Resembling a drupe in general appearance but not necessarily with its actual structure.

drupe.   A simple, fleshy fruit, derived from a single carpel, consisting of three layers; an outer skin, the exocarp, an inner fleshy mesocarp, and a hard, stony or woody endocarp which develops from the inner layer of the ovary wall and encloses the solitary seed, e.g. peaches, olives, cherries. *Syn.* stone fruit. (*See* fig. D-6).

drupecetum.   An aggregation of drupelets.

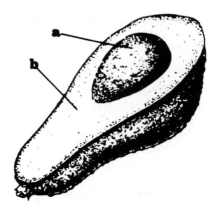

Fig. D-6. Avocado drupe: a. seed, b. fleshy pericarp.

drupelets.   One drupe of a fruit composed of aggregate drupes, as in blackberries and raspberries. Diminutive of drupe.

druse.   A globular or spherical cluster of crystals, with many secondary or component crystals projecting from its surface. May be attached to a cell wall, or lying free in the cell.

dry fruit.   Any fruit formed from an ovary with walls that are dry at maturity.

dry weight.   Moisture-free weight, obtained by drying a sample or specimen at high temperatures.

duct.   An elongated tube or canal which carries or conducts various substances such as resins, latex and oils; formed by separation of cells from one another and/or by dissolution of cells.

duff.   The partially decomposed organic matter (e.g. leaves, flowers, fruits, etc.) found beneath plants, as on a forest floor. *Syn.* litter.

dumetose, dumose.   Bushy, shrublike.

dunes.   Small hills or ridges formed by wind-blown sand.

duplication.   Occurrence of a segment of a chromosome, or a gene, twice in the haploid set.

dwarf male filament.   In *Oedogonium* (a green alga), the short filament of cells produced by the androspore, some of the cells of which are antheridia and produce male gametes.

dyad.   Pollen grains occurring in clusters of two, or in pairs.

dysgenics.   Retrogressive evolution caused by a disproportionately large reproductive success by genetically inferior individuals in a population.

dysploid.   1. In a series of related taxa, having chromosome numbers in a series but not in a polyploid sequence; i.e. not with an exact series of multiples from a base number. **2.** An individual cell or plant whose chromosome number departs markedly from normal but is not polyploid.

dystrophic lake.   A lake of low productivity which has a high content of organic matter and a low amount of dissolved oxygen; e.g. brown water, humic, and bog lakes.

dystropic, dystropous.   Referring to insects which show no special adaptation for the flowers they visit, and are frequently destructive to them, but may cause pollination.

E

e-, ex-. A prefix meaning without, out of, from.

early wood. Xylem which makes up the first wood of a given growing season. It is less dense and has larger cells than wood formed later (late wood). Replaces the term spring wood.

ebracteate. Without bracts.

ecad. A form showing adaptation to a particular habitat in which the adaptive characters are not genetically determined.

echinate. Covered with spines, prickly.

echinulate. Covered with small, pointed spines.

eco-. A prefix meaning house or home.

ecocline. A variational trend correlated with an ecological gradient.

ecodeme. A deme occurring in a specified kind of habitat.

ecological isolation. *See* isolation.

ecological niche. *See* niche.

ecology. The study of the relationships between organisms and their environment.

economic botany. The study of plants of importance to man, especially those of agriculture and industry.

ecophene. A characteristic phenotype produced as a result of environmental conditions; the resulting modifications are essentially reversible. *Syn.* ecophenotype.

ecospecies. All individuals (ecotypes), so related that they are able to exchange genes freely without loss of fertility or vigor in the offspring. *Syn.* hologamodeme.

ecosphere. *See* biosphere.

ecostate. Leaves which lack a costa or midrib.

ecosystem.   An interacting system of living organisms and their physical environment which is independent of other groups and self-sustaining, provided it receives radiant energy.

ecotone.   A transition zone; a region of overlapping plant associations, as that between two biomes or two adjacent ecosystems.

ecotype, ecotypic.   A subgroup of a species, which is genetically adapted to a particular environment; it normally has a large geographical distribution.

ectal excipulum.   *See* excipulum.

ecto-.   A prefix meaning outside, out or outer.

ectocarp.   *See* exocarp.

ectodesma, *pl.* **ectodesmata.**   *See* teichode.

ectohydric.   Land plants that transport water over their external surfaces by capillary action.

ectomycorrhiza.   One of two principle forms of mycorrhiza, in which the fungus envelops the root tip with a dense sheath, and penetrates into intercellular spaces of the host; such roots are short, branched, and appear swollen. Compare endomycorrhiza. (*See* fig. M-11).

ectopholic.   Having phloem external to the xylem.

ectophloic siphonostele.   A stele in which the phloem occurs only on the outer side of the xylem cylinder; most commonly found in seed plants.

ectoplast.   *See* plasma membrane.

ectosexine.   The outer part of the sexine of a pollen grain.

edaphic.   Pertaining to, or influenced by, soil conditions.

effective population size.   The average number of individuals in a population which contribute genetic material to the next generation.

efflorescence.   The period of flowering, anthesis.

effuse, effused.   Expanded, spreading; e.g. as a basidiocarp that is spread out or flattened.

egg.   A nonmotile female gamete. *Syn.* oosphere, ovum, macrogamete and megagamete. (*See* figs. A-18, E-2, O-3).

egg apparatus.   Normally a group of three cells, egg and two synergids, located at the micropylar end of the female gametophyte (or embryo sac) of angiosperms.

ejectosomes.   A cytoplasmic organelle ejected when the organism is disturbed. Found in some yellow-green algae (Chrysophyta). Also called trichocysts.

ektexine.   The sculptured outer layer of the exine of a pollen grain. *Syn.* sexine. (*See* fig. S-6).

elaioplast.   A type of leucoplast in which oil is formed and stored.

elaiosome.   A seed or fruit outgrowth in which oil is stored. Serves as food for ants.

elater.   **1.** An elongated, spindle-shaped, sterile hygroscopic cell in the sporangium of a liverwort sporophyte, capable of movement in response to changes in humidity, thereby aiding spore dispersal. **2.** One of four elongate, club-shaped hygroscopic appendages attached to various spores, e.g. horsetails (*Equisetum*). (*See* fig. E-1).

electrolyte.   A substance that dissociates into ions in an aqueous solution and makes possible the conduction of an electric current through the solution.

electron.   A subatomic particle with a negative electrical charge equal in magnitude to the positive charge of the proton. Electrons surround the positively charged nucleus of the atom, and determine its chemical properties. Its mass is 0.000549 atomic mass units.

electron carrier.   An enzyme which can gain or lose electrons, e.g. flavoprotein or cytochrome.

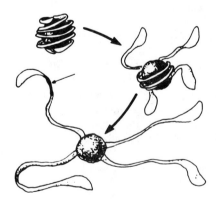

Fig. E-1. Elaters.

electron transport chain.   A chain of molecules found in mitochondria which function as hydrogen and electron acceptors. The energy released is used to transform ADP to ATP.

electrophoresis.   The differential movement of charged molecules in solution through a porus medium in an electric field. The porous supporting medium may be filter paper, cellulose acetate or a gel.

elementary particles.   *See* oxysomes.

elliptic, elliptical.   In the form of a flattened circle more than twice as long as broad.

elongate.   Lengthened; stretched out.

elongation region.   One of the four generalized regions of growth in a young root in which the cell walls increase in length, new protoplasm is formed, and vacuoles increase in size.

emarginate.   Bearing a shallow notch at the apex.

emasculate, emasculation.   Removal of anthers from a bud or flower.

embracing.   Encircling, enclosing, clasping by the base; amplectant.

embryo.   A young sporophyte while still retained within the gametophyte or seed; the minute plant produced usually as a result of fertilization and development of the zygote; in a seed, usually consists of epicotyl, hypocotyl, radicle and one or more cotyledons. (*See* fig. A-9).

embryogeny, embryogenesis.   Development of the embryo.

embryoid.   An embryo developed from a totipotent cell in tissue culture.

embryology.   The science which deals with the early development of an organism.

embryonal tube.   Secondary suspensor-like cells in the proembryos of certain gymnosperms and other vascular plants. *Syn.* secondary suspensor. (*See* fig. S-4).

embryo sac.   The female gametophyte of angiosperms. The cellular arrangement found in 70 percent of the known angiosperms consists of an eight nucelate, seven-celled structure. The seven cells are the egg cell (ovum) and two synergids at the micropylar end, three antipodal cells at the opposite end, and a central cell composed of two polar nuclei. (*See* fig. E-2).

emergences.   Plant outgrowths formed from either epidermal (e.g. prickles) or subepidermal tissues.

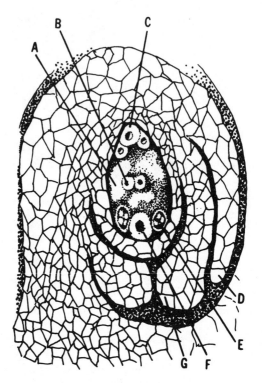

Fig. E-2. Embryo sac: A. embryo sac, B. polar nuclei, C. antipodals, D. outer integument, E. synergid, F. egg, G. inner integument.

emergent, emersed.   Having part of a plant aerial and the rest submersed; with parts extending out of the water.

enation.   An epidermal outgrowth.

enation theory.   A theory that regards microphylls as simple enations in contrast to megaphylls, which are considered to have evolved from branch systems.

encyst.   1. To form a cyst. **2.** To become enclosed within a cyst.

endarch.   Differentiation of primary xylem centrifugally with the oldest (protoxylem) closest to the center of the axis. Typically found in stems of seed plants.

endemic.   Native or confined naturally to a particular and usually very restricted geographical area or region. Compare indigen, native.

endergonic.   Energy requiring, as in some chemical reactions, e.g. photosynthesis. Contrast exergonic.

endexine.   The inner nonsculptured layer of the exine of a pollen grain. *Syn.* nexine.

endo-.   A prefix meaning within or inner.

endobiotic.   An organism which lives within another, usually in the cells of its host.

endocarp.   The innermost differentiated layer of the pericarp or fruit wall; e.g. the stony part of a drupe or pome.

endoconidium, *pl.* **endoconidia.**   An asexual spore formed inside a hypha (endogenously), and extruded.

endodermis.   A layer of cells forming a sheath around the vascular region in roots, rhizomes and some stems; by convention, it delimits the innermost layer of the cortex. Usually characterized by the presence of cells which have Casparian strips within radial and transverse walls. (*See* figs. C-3, C-4, H-6, M-2).

endodermoid.   Resembling the endodermis.

endogamy.   Inbreeding between closely related individuals. Continued endogamy results in increased homozygosity.

endogenous.   Originating from deep-seated tissues, e.g. as in lateral roots.

endohydric.   Descriptive of land plants that transport water internally through a system of special conducting cells.

endomembrane system.   A collective term for the membranes found inside a cell.

endomitosis.   The process of doubling the basic set of chromosomes in a cell resulting in formation of exact duplicates of the basic set; e.g. 1n to 2n, or 2n to 4n. Frequently occurs in cells undergoing plant tissue culture.

endomycorrhiza.   One of two principle forms of mycorrhiza, in which the fungus is not conspicuous on the root tip; it develops intracellular hyphal infections in the host root cortex; such roots are not markedly affected in form but are darker in color; the most common type of mycorrhiza. Compare ectomycorrhiza. (*See* fig. M-11).

endophyte, entophyte.   A plant growing within the tissues of another plant. (*See* fig. E-3).

endoplasmic reticulum (ER).   An extensive system of double membranes present in

Fig. E-3. Endophyte: A. plant in the family Rafflesiaceae with only flowers appearing outside host plant tissues.

most cells; it ramifies through the cytoplasm, dividing it into compartments or channels. ER may have ribosomes attached to it (rough ER) or not (smooth ER).

endopolyploidy.   Polyploidy resulting from nuclear division which is not followed by cell division; frequently found in meristematic cells.

endoscopic.   Referring to the type of embryo polarity in which the apical pole of the embryo is directed towards the base of the archegonium. It is the most common type of polarity; occurs in many lower vascular plants and in all seed plants. *Opp.* exoscopic.

endosperm.   **1.** In angiosperms, an embryonic nutritive tissue formed during double fertilization by the fusion of a sperm with the polar nuclei. **2.** In gymnosperms, a food reserve derived from the megagametophyte. (*See* fig. R-1).

endosperm nucleus.   **1.** The nucleus that results from fusion of the polar nuclei and a male gamete (sperm). **2.** The male gamete (sperm) that fertilizes the polar nuclei to form endosperm.

endosporal, endosporic.   The development of a gametophyte within the confines of the spore wall; characteristic of angiosperms.

endospore.   **1.** A spore produced within a parent cell; in bacteria, a thick-walled resistant spore; in the blue-greens, a thin-walled spore. **2.** An inner layer of the megaspore wall of *Selaginella.*

endostome.   The inner portion of the peristome.

endosymbiosis.   A symbiotic relationship between two organisms, one of which lives within the other.

endothecium. **1.** The internal tissue of an embryonic sporophyte. **2.** The inner portion of an anther wall, usually with secondary wall thickenings. (*See* fig. T-1).

endothelium. *See* integumentary tapetum.

endotoxins. Toxins retained within a bacterial cell until released by disruption of the cell.

endozoic. Living in an animal.

energy. The potential for doing work. Various forms of energy are radiant, heat, electrical, chemical and kinetic.

energy flow. The transformation of solar energy by photosynthetic organisms into chemical energy and the passage of that energy through various trophic levels in an ecosystem.

enphytotic. A plant disease which causes constant and recurring damage from year to year.

ensiform, ensate. Sword-shaped.

entire. Without indentations or incisions on the margin; smooth.

entomophilous, entomophily, entomogamy. Pollination by insects. Contrast anemophilous.

enucleate. Without a nucleus.

environment. The sum total of the external conditions which affect growth and development of an organism.

enzyme. A complex protein catalyst produced in living cells which, even in very low concentration, speeds up the rate of certain chemical reactions but is not consumed in the reaction.

eophyll. One of the first few transitional type expanded leaves, produced by a seedling before the formation of adult leaves.

epappose. Without a pappus.

epetiolate, epetiolulate. A leaf or leaflet without a petiole; sessile.

ephemeral. Lasting for only a short time, often less than one day.

epi-. A prefix meaning upon, over or beside.

epibasidium, *pl.* **epibasidia.** The upper portion of the basidium of the heterobasidiomycetes.

epibiotic. **1.** An organism that lives on the surface of another organism. **2.** A near-endemic species that represents a relic from a past flora.

epiblast. A small appendage present opposite the scutellum in the embryo of some grasses, which lacks vascular tissue. Sometimes considered to be a rudimentary cotyledon; also thought to be an artifact of median longitudinal sections. (*See* fig. R-1).

epiblem. Epidermis of a root.

epicalyx. A whorl of bracts adjacent to and resembling a true calyx.

epicarp. *See* exocarp.

epicone. The portion of the body anterior to the annulus, in a dinoflagellate.

epicormic shoots. Adventitious shoots that develop from dormant lateral buds on a tree trunk.

epicotyl. The portion of the axis of an embryo or young seedling above the point where the cotyledon(s) is attached.

epidermis. The outer layer of cells of the primary tissues of a plant, e.g. of leaves, young stems, and young roots. Usually consists of one layer, but sometimes more (multiseriate epidermis). (*See* figs. C-4, H-6, M-2, M-4, T-1).

epigeal, epigean. Describing seed germination in which the cotyledons emerge from the seed and are elevated above the soil surface. Compare hypogeal.

epigenous. Growing or developing on or above the soil surface.

epigynous, epigyny. Borne on or arising from an ovary; describes a flower which has an inferior ovary, with the sepals, petals and stamens attached at or above the top of the ovary. (*See* fig. E-4).

epimatium. A fleshy outgrowth covering the ovule and more or less fused with the integument.

epinasty. **1.** The twisting and bending of stems which occurs under certain conditions. **2.** The downward bending of leaves, especially in response from exposure to ethylene. Compare hyponasty.

epiparasite. *See* hyperparasite.

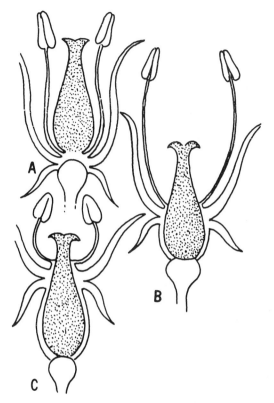

Fig. E-4. Epigynous: A. hypogynous or superior ovary, B. perigynous, C. epigynous or inferior ovary.

epipetalous.   Having stamens attached to or inserted upon the petals or corolla.

epipetric.   Growing upon rock.

epiphragm.   A multicellular membrane which closes the mouth of the capsule in certain mosses.

epiphyll, epiphyllous.   Growing upon or emerging from a leaf.

epiphyte.   An organism that grows on another plant but is not parasitic on it.

epiphytotic.   The sudden and widespread development of a destructive plant disease.

epiplasm.   The cytoplasm remaining in an ascus after spore formation; residual cytoplasm remaining following free cell formation.

episepalous.   Having stamens attached or inserted upon sepals or calyx.

episome.   In bacteria, a unit of genetic material in the cell that can either exist independently of, or be integrated into the bacterial chromosome.

epistasis, epistatic.   The condition where one gene controls, or dominates the expression of a nonallelic gene; also used to describe all types of interallelic interactions between nonallelic loci.

epistomatic stoma.   Referring to leaves in which stomata occur only on the upper surface.

epitheca.   The larger outer half of the frustule of diatoms.

epithecium, *pl.* **epithecia.**   A thin layer of tissue on the surface of the hymenium of an apothecium, formed from the tips of the paraphyses.

epithelial cell, epithelium.   A specialized cell or cells lining internal cavities within a plant, as in the resin ducts of conifers. They usually secrete various aromatic substances.

epithem.   Mesophyll of a hydathode which functions in secretion of water.

epithet.   A single descriptive word applied to species; makes up the second component of the scientific name.

epitropous ovule.   A pendulous or hanging ovule with micropyle above, and the raphe either dorsal or ventral.

equational division.   A division of a chromosome into two equal longitudinal halves (chromatids); characteristic of mitosis.

equatorial plate.   A region or plane (not a structure) passed through the center of the cell, along which the chromosomes are aligned during metaphase.

equinoctial.   Plants with flowers which open and close regularly at given hours of the day.

equisetoid hairs.   Hairs having the appearance of miniature *Equisetum* plants, found in several fern genera.

equitant.   Leaves two-ranked with overlapping bases, usually sharply folded lengthwise along midrib.

ER.   An abbreviation for endoplasmic reticulum.

era.   A major division of geologic time.

eramous.   Having unbranched stems.

erect.   Upright in relation to soil surface; perpendicular to axis of attachment as in certain flowers.

erectopatent.   Intermediate between spreading and erect.

eremad, eremophyte.   A desert plant.

ergastic substances.   Various food reserve and waste materials found in plant cells such as starch grains, fat droplets, crystals and fluids; occur in cytoplasm, vacuoles and cell walls.

ergot.   A fungal disease of grass ovaries caused by *Claviceps purpurea,* an Ascomycete (sac fungus); produces a variety of alkaloids that are poisonous and/or medically important; e.g. lysergic acid diethylamide (LSD), a psychedelic drug.

erose, eroded.   A margin which is irregularly eroded, gnawed or jagged.

erosion.   The removal of soil and other materials by wind, moving water, or by other means.

error variance.   Variance arising from unrecognized or uncontrolled factors in an experiment; it is compared with the variance of recognized factors in tests of significance.

eseptate.   *See* aseptate.

espalier.   A framework or trellis upon which plants are trained to grow; or the plant that is so trained.

essential elements.   Elements required by plants for normal growth and development.

essential oils.   A class of aromatic, highly volatile substances which occur in many types of plants. These oils are produced in a variety of plant organs, e.g. petals, leaves and bark.

essential organs.   The gynoecium and androecium.

estipitate.   Without a stipe.

estipulate.   Without stipules.

estivation.   The arrangement of floral parts within a floral bud; also aestivation.

estuary.   An inlet of the sea, especially at the mouth of a river where the tide meets the current of the river.

ethnobotany.   The study of the relationship of man, especially (but not limited to) primitive man, and his use of the surrounding vegetation.

ethological isolation.   Reproductive isolation due to pollinator behavior.

ethylene.   A gaseous hormone that effects numerous physiological reactions in plants; e.g. it promotes fruit ripening, inhibits elongation of cells, induces flowering (in pine-apple), etc.

etiolated.   Plants which have elongated stems, poor leaf development and lack chlorophyll as a result of being grown in insufficient light.

eucarpic.   A condition in fungi in which only part of the thallus, rather than the entire thallus, is converted into reproductive structures.

euchromatin.   Chromatin in chromosomes which is dispersed between mitotic and meiotic cell divisions; it is condensed during nuclear divisions into chromosomes and stains darkly. Considered to be the genetically active portion of chromosomes, or the active state of a particular gene. Compare heterochromatin.

eukaryotic, eucaryotic.   Organisms which have true nuclei, and other membrane-bound organelles, and have chromosomes, plastids, ER, etc. Contrast prokaryotic.

eumeristem.   "True meristem", containing relatively small, thin walled, compactly arranged, isodiametric cells, which have dense cytoplasm and large nuclei.

euphilic.   A flowering plant which is specially adapted for pollination by a few specialized vectors.

euphotic zone.   The upper layer of a body of water in which light is sufficient for photosynthesis. Average thickness in the sea is 100 fathoms (about 600 feet). (*See* fig. A-16).

euploid.   A polyploid with a chromosome number that is an exact multiple of the base number of the polyploid series.

euryplastic.   A genotype with a wide range of plasticity. *Opp.* stenoplastic.

eurytopic.   An organism found in many different habitats.

eusporangiate.   Sporangia of ferns and lower vascular plants whose walls develop from several superficial leaf cells. Sporangia have several wall layers and are usually large with numerous spores.

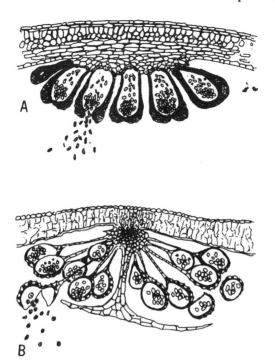

Fig. E-5. Eusporangium: A. eusporangium, B. leptosporangium.

eusporangium, *pl.* **eusporangia.**   A sporangium which has several wall layers, is usually large and has numerous spores; develops from several superficial leaf cells. Compare leptosporangium. (*See* fig. E-5).

eustele.   A stele characteristic of dicots and gymnosperms in which the primary vascular tissues are arranged in a cylinder of anastomosing vascular bundles around the pith. The phloem is only found outside of the xylem.

eutrophic, eutrophication.   Descriptive of a body of water which is rich in organic and inorganic nutrients.

eutropic.   Animal vectors that are specialized for getting most of their food from flowers.

evagination.   An outgrowth, unsheathing or outpocketing.

evanescent.   Disappearing quickly.

evapotranspiration.   Total water loss from an area by evaporation and transpiration.

evapotranspiration gauge. *See* lysimeter.

evergreen. Plants which have persistent green leaves for two or more growing seasons; leaves of evergreens are usually shed over a long period of time, in contrast to leaves of deciduous plants which are shed annually. Compare deciduous.

evolution. Genetically fixed change through time, particularly with reference to organisms.

ex-. A prefix meaning out, off, from or beyond.

exalbuminous. Descriptive of seeds that lack endosperm.

exannulate. Lacking an annulus.

exarch. Differentiation of primary xylem centripetally, with the oldest xylem (protoxylem) located nearest the outer portion of the axis.

excentric. One-sided; off-center.

excipulum, *pl.* **excipula.** A specialized, fleshy portion of an apothecium supporting the hypothecium and hymenium. Consists of two layers, the ectal excipulum or outer layer, and the medullary excipulum or inner layer.

excise. To cut out or remove a part.

exclusion principle. States that no two species can coexist in the same niche if they have identical ecological requirements.

excoemum. A tuft or fringe of pubescence at the base of the glumes in certain grasses.

excretion. The elimination of the end products of metabolism. *See* secretion.

excurrent. **1.** Extending beyond the apex or margin of a leaf into a mucro or awn. **2.** Descriptive of tree growth in which the main axis continues to the top of the tree, from which smaller, lateral branches arise.

exergonic. An energy releasing chemical reaction. Contrast endergonic.

exfoliate, exfoliating, exfoliation. Peeling off in shreds, thin layers, sheets or plates, as in the bark or epidermis of certain trees and shrubs.

exindusiate. Lacking an indusium.

exine. The outer wall layer of spores and pollen grains, usually divided into two main layers: an outer ektexine and an inner endexine. (*See* fig. S-6).

exo-. A prefix meaning out, outside or without.

exobiology. The study of extraterrestrial life.

exocarp. The outermost layer of the pericarp or fruit wall; often the mere skin of the fruit. Also called epicarp or ectocarp.

exoconidium, *pl.* **exoconidia.** An asexual spore formed on the surface or at the tip of a hypha.

exodermis. A type of hypodermis which is structurally similar to the endodermis. It is the outermost layer of the cortex of some roots and may be one or more cells thick.

exoenzyme. An enzyme formed or secreted outside the protoplast of a cell and functioning outside of it.

exogamy. Cross-fertilization or the union of gametes with different genetic makeup; outbreeding. Exogamy increases heterozygosity.

exogenous. Produced externally or developing on the outside.

exoscopic. Referring to the type of embryo polarity in which the apex (apical pole) of the embryo faces outward toward the neck of the archegonium, as in horsetails (*Equisetum*). *Opp.* endoscopic.

exosporal. With the gametophyte developing outside the confines of the spore wall.

exospore. A spore that is not produced within a sporangium.

exostome. The outer portion of the peristome.

exothecial cells. The epidermal cells of a moss capsule.

exotic. Foreign in origin, not native.

exotoxins. Soluble toxins produced by plants that diffuse out of a living cell into the environment.

explanate. Spread out flat.

explant. An excised tissue or organ fragment which is used to initiate an *in vitro* culture.

explosive fruit. One which bursts suddenly and usually violently, scattering seeds over a considerable area. (*See* fig. E-6).

Fig. E-6. Explosive fruit.

explosive hairs.   In many Cucurbitaceae, hairs on the floral organs that contract force-fully when the easily detached tips are touched, expelling the contents on the touching object. In some cases (e.g. *Cyclanthera pedata*) the protoplasmic substance quickly hardens, and might immobilize small insects; in other cases the trichome contents may moisten the legs and bodies of visiting insects and aid pollen adhesion.

explosive speciation.   The rather rapid production of several new species from a single species in a given geographical region.

expressivity.   The degree of manifestation of a genetic character; the manner in which a phenotype is expressed.

exserted.   Projecting beyond, sticking out or protruding.

exstipulate.   Without stipules.

extant.   Presently living or existing.

extinct, extinction.   No longer living.

extra-.   A prefix meaning outside of or beyond.

extrafloral nectaries.   Nectaries located outside the flower, as in leaves or bracts. (*See* fig. E-7).

extraxylary fibers.   Fibers found outside the xylem. Compare xylary fibers.

extrazonal vegetation.   Vegetation which extends out of its normal zone or range and is able to exist there because of the mediating effect of slope exposure; e.g. warm zone vegetation occurring in cooler zones on southern exposures.

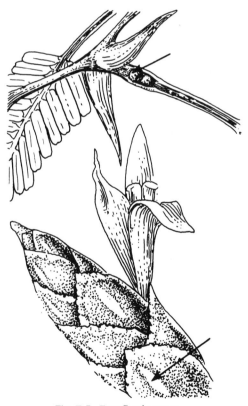

Fig. E-7. Extrafloral nectary.

extrorse.   Opening or facing outward; as with anthers that dehisce toward the outside of the flower.

exudate, exudation.   Fluids passed out through small openings in a plant.

exude.   To discharge slowly through small pores.

eye.   **1.** The center of a flower, often uniquely colored. **2.** A bud of an underground storage stem, e.g. a tuber (*See* fig. T-8).

eyespot.   A small pigmented area often lying near the base of the flagella in certain unicellular organisms, e.g. euglenids and some algae. It may be inside or outside a plastid and may be light sensitive.

F

F₁. The first filial generation (hybrid) in a cross between two parental types.

F₂. The second filial generation resulting from the self-fertilization of F_1 individuals.

facultative. Descriptive of an organism which has the ability to live under different nutritional conditions, depending on the environment; e.g. a facultative parasite or a facultative saprobe.

fairy ring. A circular growth pattern on the ground representing the appearance of the fruiting bodies of mushrooms (Basidiomycetes).

falcate. Sickle-shaped. *Syn.* seculate.

falcate-secund. Leaves falcate and all turned toward one side of stem.

fallow. Agricultural land left uncultivated for one or more seasons to allow for the accumulation of moisture, destruction of weeds, and the decomposition of organic matter.

falls. The normally drooping portion of the perianth of an *Iris* flower. Compare standard.

false annual ring, false growth ring. A growth ring in wood, as seen in cross section, formed from more than one growth layer during a single growing season.

false branching. The breakage of a filament resulting in one or both ends protruding from the sheath. Found in some blue-greens (Cyanophyta).

false indusium. In ferns, a covering formed by the inrolling of a leaf margin over marginal sporangia. (*See* fig. F-1).

family. A taxonomic unit above the genus and below the order, composed of one or more genera. Plant families end in the suffix -aceae; animal families and some protista end in -idae.

farinaceous, farinose. Covered with a mealy power; mealy in texture; containing starch or starchlike materials. *Syn.* mealy.

farming. The science and art of crop production. The direct application of the multiple aspects of agronomy.

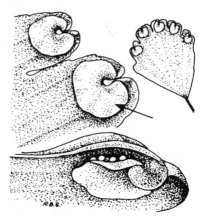

Fig. F-1. False indusium on abaxial surface.

fasciated, fasciation.   An unnatural or sometimes teratological widening and flattening of a stem.

fascicle, fascicled.   A close cluster or bundle of flowers, leaves, stems or roots. (*See* fig. L-6).

fascicular cambium.   A vascular cambium originating between the xylem and phloem in a primary vascular bundle.

fasciculate.   Leaves or other structures in clusters, or bundles originating from a common point.

fastigiate.   Strictly erect and more or less parallel; e.g. as in reference to branches.

fats.   Organic compounds containing carbon, hydrogen and oxygen, as in carbohydrates. The proportion of oxygen to carbon is much less in fats than it is in carbohydrates. Fats in the liquid state are called oils. *See* lipid.

fatty acid.   An organic acid composed of a long chain of hydrocarbons; three such acid molecules joined to a glycerol molecule form a fat molecule.

faucal, fauces, faucial.   Referring to the throat area of a gamopetalous corolla.

fauna.   A collective term for animal life.

faveolate, favose.   Honeycombed, as the receptacle in many species of the sunflower family (Asteraceae).

fecundity.   The capacity of producing abundant seeds or offspring; fertility.

feedback.   The influence that the result of a process has on the functioning of that process.

felted.   Closely matted with intertwined hairs.

felt hairs.   Dense trichomes produced early in the development of a leaf, floral part, etc., as a protective covering.

female.   Descriptive of a plant or plant part which produces the ovum or female gamete; pistil. Symbol ♀.

female symbol ♀.   The looking glass, the zodiac sign for Venus.

fenestrate.   Perforated with windowlike holes or translucent areas.

feral.   Wild; the reversion of a cultivated plant to a wild state.

fermentation.   A respiratory process occurring in the absence of, or with only limited oxygen. Substantially less ATP is produced by this process than is produced in aerobic respiration; the end product may be ethyl alcohol, lactic acid or other products.

ferredoxin.   An iron-rich protein involved in electron transfer in photosynthesis.

ferruginous.   Rust-colored.

fertile.   Capable of reproducing; used in reference to an entire organism, or to part of it (e.g. stamens or carpels), or to seed-bearing fruits.

fertilization.   1. The union or fusion of two gametes to produce a zygote. 2. The process of adding soil amendments to increase fertility.

fertilization chamber.   An archegonial chamber.

fertilization tube.   A tube produced in some water molds (Phycomycetes) by which the male gametangium transfers male gametes to the female gametangium. (*See* fig. F-2).

fertilizer.   Any substance added to the soil which provides elements essential to plant growth, or brings about a balance in the ratio of nutrients in the soil.

fetid.   Having a disagreeable odor.

fiber.   1. An elongated, generally thick walled, often lignified sclerenchyma cell which tapers at the ends. It may or may not be living at maturity. 2. Any of a number of elongated structures which are used by the textile industry and are obtained from plants, e.g. cotton. Compare sclereid.

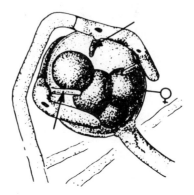

Fig. F-2. Fertilization tube.

fiber-sclereid.   A sclerenchyma cell morphologically intermediate between a fiber and a sclereid.

fiber-tracheid.   A fiberlike tracheid in the secondary xylem; commonly thick-walled, with pointed ends and bordered pits that have lenticular to slitlike apertures.

Fibonacci phyllotaxis or series.   A method for analyzing leaf arrangements by expressing mathematically, as a fraction, the angle of divergence between two successive leaves; e.g. ½, ¹/₃, ²/₅ etc. in which each succeeding fraction is the sum of the two previous numerators and the sum of the two previous denominators.

fibril.   Submicroscopic threads or fibers that constitute the form in which cellulose occurs in the cell wall. *See* microfibril and macrofibril.

fibrillose.   Covered with little fibers.

fibrous.   Having fine or slender structures.

fibrous root system.   A root system composed of many roots similar in length and thickness; as in grasses and many other monocotyledons. *Syn.* diffuse root system.

fibrovascular.   Composed of conducting cells, e.g. xylem vessels and sieve tubes, and their fibrous sheaths.

-fid.   A suffix meaning deeply cut or cleft.

fidelity.   1. The degree to which a species is restricted to a particular type of community. 2. The degree to which an individual pollinator is specific in its visitations to a single species of plant on a given flight.

field capacity, field percentage. The amount of water a particular soil will hold against the action of gravity.

filament. **1.** A threadlike process, structure or growth form. **2.** The stalk of a stamen which supports the anther.

filamentous. With a threadlike cellular growth form; elongate, formed of filaments or fibers.

file meristem. *See* rib meristem.

filial. Of or pertaining to offspring. *See* F_1.

filical, filicoid. Fernlike.

filiform. Threadlike, usually flexuous.

filiform apparatus. A complex of cell wall invaginations in a synergid cell similar to those in transfer cells; thought to increase the absorbing capacity of the cell.

filiform sclereid. A slender, much elongated sclereid, resembling a fiber. (*See* fig. S-3).

filling tissue. The loose tissue formed outwardly by the lenticel phellogen; may or may not be suberized. *Syn.* complementary tissue.

fimbria. **1.** A proteinaceous bacterial tube through which DNA passes during conjugation. **2.** A fringe or fringelike border.

fimbriate. Margins fringed, with long and coarse hairs.

fimbrilla. A single trichome from a fringe of hairs.

fimbriolate. Minutely fimbriate.

fission. The asexual duplication of a cell or of a one-celled organism by the division of the cell or organism into two, more or less equal parts.

fistular, fistulose. Hollow and cylindrical; as the leaf and stem of an onion.

fitness. The relative ability of a genotype (individual) to survive and produce viable and fertile offspring; the reproductive capacity of an organism. *Syn.* adaptive value.

fixation. The initial step in the procedure of making permanent preparations for microscopic study. Involves the killing and preventing of decay of the specimen.

flabellate, flabelliform.   Fan-shaped; broadly wedge-shaped.

flaccid.   Weak, limp, lax or flabby.

flag.   A branch with dead leaves on an otherwise living tree.

flagellate.   Referring to cells that possess special organelles for motility.

flagellates.   Unicellular eukaryotic organisms that move through water by means of flagella.

flagelliform.   Whip-shaped.

flagellin.   The proteinaceous component of flagella.

flagellum, *pl.* **flagella.**   A hair-, whip- or tinsellike elongated structure which serves to propel a motile cell.

flavedo.   The yellow-colored tissue of the citrus rind.

flavescent.   Yellowish.

flavoprotein.   A yellow, dehydrogenase enzyme which is important in cellular oxidation.

fleshy.   Succulent, thick and juicy.

flexuous.   Coarsely sinuous or wavy; curved alternately in opposite directions.

floccose, flocculose.   Bearing dense, appressed trichomes in patches or tufts; woolly.

flocculent.   The diminutive of floccose.

flora.   1. The plant life of an area or period. 2. A work which contains the description and classification of the plants of a particular region.

floral.   Of or pertaining to flowers.

floral diagram.   A diagram showing the relative position and number of the floral parts.

floral envelope.   The perianth, or calyx and corolla.

floral formula.   An abbreviated method of expressing the information illustrated in a floral diagram, e.g. number, sex, shape, etc. of the floral parts.

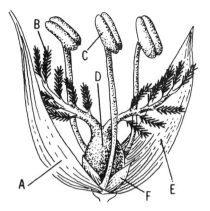

Fig. F-3. Grass floret: A. lemma, B. stigma, C. anther, D. ovary, E. palea, F. lodicule.

floral tube.   A tube or cup formed by the fused basal portions of the sepals, petals and stamens that surround the ovary (fused or free from it); often in perigynous and epigynous flowers.

florescence.   Anthesis, the time or period of flowering.

floret.   Small individual flowers that make up a very dense inflorescence, as those in the head of a composite flower, or the spikelet of a grass spike. (*See* fig. F-3).

floricane.   A flowering and fruiting stem, especially of brambles.

floriculture.   The cultivation of flowers for ornamental purposes.

floridean starch.   The reserve carbohydrate formed by the red algae (Rhodophyta).

floriferous.   Bearing or producing flowers.

florigen.   The postulated hormone or group of hormones that are thought to induce flowering.

flower.   The specialized reproductive structure of the angiosperms.

flower bud.   A bud which develops into one or more flowers.

fluke cell.   Any of the lateral, projecting cells of an anchor hair.

fluorescence.   A reemission of absorbed radiation at a longer wavelength than the original.

flypaper trap.   In carnivorous plants, a trap on which the victims are ensnared by a sticky, mucilagenous secretion.

foliaceous.   Leaflike; resembling a foliage leaf in shape, texture, size and/or color.

foliage.   A collective term for the leaves of a plant.

foliar.   Pertaining to leaves or leaflike parts.

foliate, foliferous.   Bearing or possessing leaves or leaflike structures.

foliolate.   Having leaflets.

foliose.   **1.** Leaflike or flattened as in the erect, bladelike thallus of certain algae. **2.** In lichens, a growth form in which the flattened thallus is attached only loosely to the substrate, and forms a broad, leaflike, often ribbed structure. (*See* fig. L-4).

follicetum.   An aggregation of follicles.

follicle.   **1.** A dry, dehiscent fruit derived from one carpel that splits along one suture. (*See* fig. A-7). **2.** Any small saclike cavity.

food chain.   The pathway of chemical energy in any natural community from producers to consumers to decomposers.

food web.   The complex interrelationships apparent in the energy flow through the interconnecting food chains of a community.

foot.   The basal portion of the embryonic sporophytes in liverworts, mosses, and certain vascular plants; it remains in contact with the gametophyte, absorbing food from it and serving as the organ of attachment.

foot candle.   A measure of illumination which represents direct illumination one foot from a uniform light source of one international candle.

forage.   Vegetative food available for domestic animals. *Syn.* fodder.

foramen, *pl.* **foramina.**   A more or less circular aperture; micropyle.

forespore.   A clear area in cytoplasm of a bacterial cell which becomes surrounded by a refractile wall in endospore formation.

forest.   A vegetation type dominated by trees.

forestry.   The science of managing forest resources, the growth and cultivation of trees.

form genera.   In fossil plants, the generic name given to isolated plant fragments which show the same form or morphology, and may or may not be naturally related. *Syn.* organ genera.

form group.   A taxon, designated on the basis of similar form or morphology and not necessarily an indication of evolutionary relationship.

formicophily.   Pollination by ants.

formula, *pl.* **formulas or formulae.**   A symbolic expression of the chemical composition of a compound.

fornix, *pl.* **fornices.**   A small arched scale or appendage.

fossil.   A natural object representing any one of several different types of plant remains, or evidence of their past existence, that are preserved in the earth's crust. (*See* fig. F-4).

foundation seed.   Seed stock produced from breeder seed by or under the direct control of an agricultural experiment station. Foundation seed is the source of certified seed.

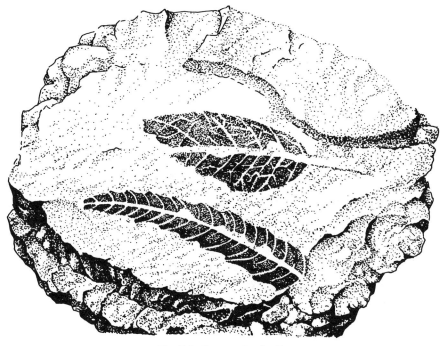

Fig. F-4. Compression fossil.

founder principle. The principle that propagules which initiate a new and isolated population will contain less genetic diversity than the parent population from which they originated.

fovea. A small depression or pit; as the depression containing the sporangium in the leaf base of *Isoetes* (quillwort).

foveolate. Pitted.

fragmentation. A method of asexual reproduction whereby an organism is broken into a number of parts, each of which is capable of producing a new organism; e.g. as in certain algae, fungi and liverworts.

free. Separate, distinct; not united in any way to other organs or structures.

free cell formation. Cytokinesis inside a preexisting cell to form one or more new cells.

free-central placentation. Attachment of ovules around a central column which is free from the compound ovary wall except at base. (*See* fig. P-10).

free energy. Energy available to do work.

free nuclear division. Nuclear division (karyokinesis) that occurs without cell wall formation (cytokinesis), resulting in a multinucleated cell; characteristic of some embryos and gametophytes.

free soil water. Water that percolates through, and is eventually lost from the soil.

freeze drying. A method of drying under a vacuum, material which has been rapidly frozen. This technique minimizes shrinkage in the specimen.

freeze etching. A technique in which living cells are very rapidly frozen and broken open to expose the inner portions of the cell. The freeze-fractured cells are then coated under a high vacuum, to form an electron reflective surface, before examination with the electron microscope.

frequency. **1.** The percentage of sample plots in which an organism occurs. **2.** The number of cycles per unit of time.

frequency distribution. The quantitative distribution of some parameter on the basis of some variable characteristics.

frequency histogram. *See* histogram.

frond. A fern or palm leaf or similar flattened leaflike structure.

fructiferous.   Bearing fruit.

fructification.   A fruit in angiosperms; a reproductive organ or fruiting body.

fruit.   In angiosperms, a ripened ovary or ovaries with or without accessory floral parts and/or seeds.

fruit wall.   *See* pericarp.

frustule.   One of the two halves of the siliceous cell wall of a diatom.

fruticose.   1. Becoming shrubby in the sense of being woody. Also called frutescent. 2. A branched, usually upright growth form of a lichen thallus; may be shrublike in appearance. (*See* fig. L-4).

fucoxanthin.   The brown pigment (xanthophyll) that is characteristic of the brown algae (Phaeophyta) and chrysophytes (Chrysophyta).

fugacious.   Ephemeral, lasting only a short time; usually applied to plant parts.

fulvous.   Tawny; dull yellowish-brown.

fumigant.   A volatile substance which kills organisms by a poisonous vapor.

functional megaspore.   The surviving megaspore of the linear tetrad in seed plants that gives rise to the female gametophyte.

fundamental tissue.   *See* ground tissue.

fundamental tissue system.   *See* ground tissue system.

fungal trap.   A specialized ring of hyphae capable of catching small animals (e.g. nematodes) that serve as food for the fungus. (*See* fig. F-5).

fungicide.   A toxic agent used to control or eradicate fungi.

Fungi Imperfecti.   An artificial group which includes a large number of fungi whose sexual stages are lacking or unknown. *Syn.* deuteromycete.

fungistatic.   An agent used to control the growth of fungi without killing them.

fungoid.   Funguslike.

fungus, *pl.* fungi.   Any of a large group of achlorophyllous, simple plants, which may be either saprophytic or parasitic. They usually reproduce by both sexual and asexual means.

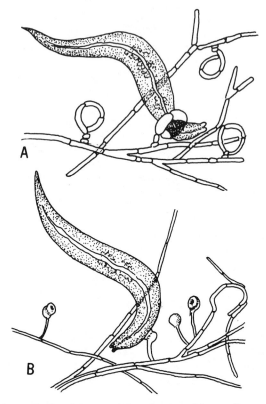

Fig. F-5. Fungal trap: A. fungal trap consisting of a ring of three cells surrounding a nematode, B. fungal trap with small adhesive knob attached to a nematode.

funiculus.   The basal stalk of an ovule arising from the placenta in the angiosperms. (*See* fig. O-3, S-7).

funnelform.   Funnel-shaped, gradually widening upwards.

furcate.   Forked.

furcellate.   Slightly furcate or forked.

furfuraceous, furfurate.   Covered with branlike scales; scurfy.

furrowed.   With longitudinal channels or grooves; sulcate.

fuscous.   Grayish-brown; dull-brown.

fusiform.   Spindle-shaped; thickest near the middle and tapering toward each end, as a fusiform cell.

fusiform initials.   One of the two types of cells that form the vascular cambium. They are meristematic, prism-shaped, fusiform cells, that give rise to elements of the axial system in the secondary phloem and secondary xylem.(*See* fig. A-23).

fusoid.   Somewhat fusiform.

G

galbulus. The modified cone of the genus *Cypress* in which the apex of each scale is enlarged and more or less fleshy.

galea. A petal shaped like a helmet; the strongly concave upper lip (petal) of some bilabiate corollas.

galeate. Helmet-shaped; having a galea.

gall. A localized abnormal growth on a plant induced by a fungus, insect or other foreign agent. (*See* fig. G-1).

gametangial contact. Sexual reproduction involving contact of gametangia followed by the transfer of the nuclei from the antheridium to the eggs through a fertilization tube, as in some water molds.

gametangial copulation. Sexual reproduction in which the entire protoplasts of two gametangia fuse; as in certain water molds.

gametangium, *pl.* **gametangia.** Any cell or organ which produces gametes.

gamete. A mature haploid reproductive cell or nucleus which unites with another gamete to form a zygote in sexual reproduction. *Syn.* sex cells.

gametocyte. A cell that gives rise to gametes by division.

gametogenesis. The process of gamete formation.

gametophore. **1.** An axis that bears gametes. **2.** In the bryophytes, a fertile stalk that bears gametangia (antheridia and archegonia).

gametophyte. The gamete-producing haploid phase in the life cycle of plants. (*See* fig. A-9).

gametothallus, *pl.* **gametothalli.** A thallus that produces gametes, as opposed to one that produces spores (sporothallus).

gamma radiation. A type of electromagnetic radiation of short wavelength; emitted from atomic nuclei undergoing radioactive decay.

gamo-. A prefix meaning united.

154

Fig. G-1. Galls.

gamodeme.   A population of individuals of a particular taxon (i.e. a deme) which are capable of interbreeding; a panmictic breeding unit.

gamone.   A hormone secreted by vegetative hyphae of certain fungi; e.g. as in the Mucorales.

gamopetalous.   Having petals which are partly or completely fused (connate). *Syn.* sympetalous.

gamophyllous.   With leaves united by their edges.

gamosepalous.   Having sepals which are partly or completely fused (connate). *Syn.* synsepalous.

gap.   **1.** A discontinuity in variation. **2.** A break in the vascular cylinder of a stem. *See* leaf gap, branch gap.

gas chromatography.   A chromatographic technique in which vapors of a substance are separated as they are swept through a column by an inert gas.

geitonogamy.   Fertilization of one flower by another flower on the same plant.

gelatinous.   Jellylike.

gelatinous fiber.   A fiber with little or no lignification and having a gelatinous appearance.

gelation.   The conversion of a liquid gel into a solid.

gel electrophoresis.   A type of chromatography in which the solutes migrate in an electrical field through a gel.

geminate.   In pairs, paired.

gemma, *pl.* **gemmae.**   **1.** A specialized group of asexual reproductive cells; an outgrowth of the thallus. They occur in certain fungi, on the gametophytes of certain liverworts, mosses and ferns, and on the sporophytes of lycopods. **2.** A process found on pollen grains; larger than one micrometer with the diameter equal to, or greater than, the height and constricted at base.

gemmae cup.   A cup-shaped structure containing gemmae, borne on the upper surface of liverwort gametophytes. (*See* fig. H-2).

gemmate.   Pollen having gemmae on their surface.

gemmiferous.   Bearing gemmae.

gemmiform.   Budlike.

gene.   A functional unit of inheritance which consists of a segment or nucleotide sequence of a DNA molecule. *Syn.* cistron.

genecology.   The study of infraspecific variability in plants in relation to environment.

gene dosage.   The number of copies of a gene found in the nucleus of a given cell.

gene flow.   The spread of a particular allele through a population as a result of outcrossing and natural selection.

gene frequency.   The proportion of a particular allele of a gene in a population.

gene interaction.   The modification of the usual expression of one gene by the action of a nonallelic gene or genes.

gene pool.   All the alleles of all genes within an interbreeding population.

generative cell.   A cell of the male gametophyte that produces the stalk and body cells in many gymnosperms; divides to form two sperm nuclei in angiosperms. (*See* fig. P-17).

gene recombination.   New gene combinations appearing in the progeny of sexually reproducing organisms as a result of the rearrangement of genetic material during meiosis.

gene substitution.   The replacement of one gene by an allele; all other relevant genes remain unchanged.

genetic, genetical.   Pertaining to or concerned with inheritance.

genetic code.   The sequence of three nucleotides in a DNA molecule that specifies the arrangement and kinds of amino acids in a polypeptide chain through RNA intermediaries (i.e. mRNA and tRNA). *See* codon.

genetic drift.   The tendency in small peripheral populations for genetic homozygosity to occur by chance; results in small, genetically stable, interbreeding groups.

genetic equilibrium.   The condition in which stability, in regard to the frequency of various genotypes, occurs in a population over successive generations.

genetic load.   The number of recessive deleterious or lethal genes present within an interbreeding population.

genetic map.   A linear arrangement of mutable sites on a chromosome; determined from genetic recombination studies.

genetic polymorphism.   *See* polymorphism.

genetics.   The study of inheritance.

gene transcription.   *See* transcription.

geniculate.   Abruptly bent at a joint, like a bent knee.

genodeme.   A deme differing from others genetically.

genome.   A complete haploid set of chromosomes, as is present in a gamete; two genomes are found in somatic cells, while more than two genomes occur in polyploids.

genophore.   The strand of DNA in prokaryotic organisms and viruses, on which the genes occur.

genotype.   The genetic constitution of an individual, in contrast to its appearance or phenotype.

genotypic variation.   Genetic variation within a population or species.

genus, *pl.* **genera.**   A taxonomic subdivision between the family and species, which includes one or more closely related species.

geocarpous, geocarpy.   Development of fruit in the ground from a flower which developed and was pollinated above ground; e.g. the peanut.

geoflora.   A biome having a large distribution in both time and space.

geoflorous, geoflory.   Having flowers blooming beneath the ground.

geographic isolation.   The separation of a population from the rest of the species, or other taxa, by a geographical barrier; geographic isolation may lead to speciation.

geographic speciation.   The process by which geographically isolated populations of a single species develop and maintain unique gene combinations and thereby become reproductively isolated. Species formation is one evolutionary result of this process.

geonastic, geonasty.   Curving toward the ground.

geophyte.   Plants with meristematic portions (buds or organs) located on the plant below the soil surface, as on bulbs or rhizomes.

geotaxis.   The orientation of an organism with respect to gravity.

geotropism.   A growth movement of a plant in response to the influence of gravity. (*See* fig. G-2).

germ.   **1.** A common name referring to the embryo, as in "wheat germ". **2.** Any microscopic disease-causing organism, e.g. bacteria.

germinal furrows, germinal pores.   Thin areas in pollen grains in the form of furrows or pores. Depending on the species, the pollen may possess either one or both types. *Syn.* germ pore.

germination.   The beginning or resumption of growth of an embryo or spore.

germ tube.   The first tubular outgrowth produced by a germinating spore.

gibberellins.   A group of plant hormones important in many physiological processes; e.g. affects cell elongation, induces parthenocarpy, promotes seed germination, induces flowering, etc.

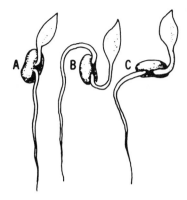

Fig. G-2. Geotropism: A. seed in normal position, B. seed upside down, C. seed on side.

gibbous.   Swollen on one side near the base. *Syn.* ventricose.

gill.   The platelike structures on the underside of the cap in the club fungi (Basidiomycetes). (*See* fig. B-2).

gill fungi.   Most mushroom members of the Basidiomycetes.

girder.   A connection of sclerenchymatous tissue between a vascular bundle and the epidermis.

girdle.   In dinoflagellates, the transverse groove which contains a flagellum. (*See* fig. D-4).

girdle band.   In diatoms, the region where the frustules overlap.

girdling.   The removal of a complete ring of bark (including the phloem) from a tree; results in death of the tree due to root starvation. *Syn.* ringing.

glabrate, glabrescent.   Nearly or becoming glabrous at maturity.

glabrous.   Without pubescence; not hairy.

glade.   An open space in a forest.

gladiate.   Sword-shaped.

gland.   A multicellular secretory structure. *See* secretory cell.

glandular, glanduliferous.   Bearing or having glands.

glandular hair. A trichome having a unicellular or multicellular head composed of secretory cells; usually borne on a stalk of nonglandular cells.

glans, *pl.* **glandes.** A nut subtended in a dry cupular involucre as in oaks (*Quercus*).

glaucescent. Somewhat or slightly glaucous.

glaucous. Covered with a removable waxy coating which gives the surface a whitish or bluish cast.

glazing. The separation of epidermis and mesophyll in leaves due to physical or chemical damage or disease. A common leaf response to air pollution.

gleba, *pl.* **glebae.** The inner, spore-producing tissue of the fruiting body of certain Basidiomycetes; e.g. Gasteromycetes.

globose, globular. Spherical or rounded.

globule. The male reproductive structure in the stoneworts (Charophyta).

glochid. A minute barbed bristle or hair, often in tufts, as in certain cacti; e.g. *Opuntia*. (*See* fig. G-3).

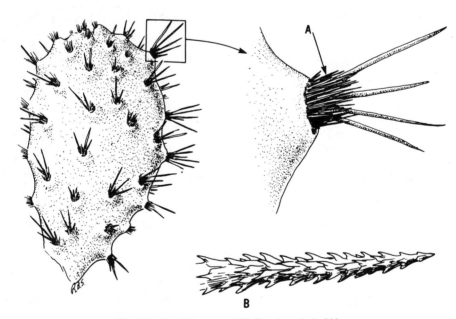

Fig. G-3. Glochid: A. glochid, B. enlarged glochid.

glochidiate.   Pubescent with trichomes barbed at the tip.

glomerate.   In dense or compact clusters or heads.

glomerule.   A small dense indeterminate cluster of more or less sessile flowers.

glucose.   A common six-carbon sugar ($C_6H_{12}O_6$); the most common monosaccharide; dextrose.

glumes.   A pair of small scalelike bracts subtending a grass spikelet.

gluten.   A "sticky" protein in the endosperm of certain grasses; allows bread dough to rise.

glutinous.   With a sticky exudate.

glycogen.   A complex carbohydrate similar to starch; a food reserve of blue-greens, fungi, and most animals. Also called cyanophycean starch.

glycolysis.   A process in which glucose is converted anaerobically into pyruvic acid with a net gain of a small amount of ATP.

glycophyte.   Plants which can tolerate only low to moderate concentrations of salt in the soil. Compare halophyte.

glyoxysome.   A microbody in many seeds containing enzymes necessary for the conversion of fats into carbohydrates during germination.

Golgi apparatus.   A collective term for all Golgi bodies in a cell.

Golgi body.   An organelle in eukaryotes consisting of an aggregation of flat or curved and hollow, disklike membranes; function in the secretion of various compounds that the cells manufacture. *Syn.* dictyosome (in plants).

gonidium.   An asexual reproductive cell.

gonimoblast.   A filament developing after fertilization in red algae (Rhodophyta) which gives rise to carposporangia.

gonophore.   A stalk that bears stamens and pistils.

G_1 phase, G_1 period.   A post-mitotic period which lasts from telophase to the initiation of the S period.

G_2 phase, G_2 period.   A pre-mitotic period which lasts from the end of the S period through the beginning of mitosis.

gradate.   An arrangement of sporangia in which the sori mature centrifugally; i.e. with the oldest at the apex and the younger ones lateral.

graft, grafting.   The transfer of aerial parts of one plant (e.g. buds, twigs, etc.) onto the root or trunk of a different plant. *See* scion, stock.

grain.   *See* caryopsis.

gram atomic weight.   The quantity of an element which has a weight in grams equal numerically to the atomic weight of the element.

gram-molecular weight.   *See* mole.

gram-negative.   Bacteria that do not stain with a Gram stain.

gram-positive.   Bacteria that stain with a Gram stain.

Gram stain.   A stain technique that involves gentian violet, an iodine solution and alcohol or acetone. Used in classifying bacteria.

granular, granulose.   Finely covered with very small mealy granules.

granum, *pl.* **grana.**   Stacks of parallel membranes (thylakoids) that occur within the chloroplasts of land plants. The grana contain chlorophyll and carotenoid pigments, and are the site of the light reactions of photosynthesis. (*See* fig. C-6).

gravitational water.   Water that is not held by the soil but passes through from the force of gravity; as opposed to capillary water which is retained in the soil.

greenhouse.   A structure in which plants are grown, frequently under conditions of controlled light, temperature, and humidity; covered with a transparent material such as glass, fiberglass or plastic.

greenhouse effect.   **1.** An increase in the temperature of the atmosphere as a result of an increased concentration of atmospheric carbon dioxide. **2.** In horticultural applications, the creation of an artificial environment surrounding plants to increase the ambient temperature.

green manure.   A green crop that is plowed under to increase the organic matter and nitrogen content of the soil.

gregarious.   Growing in groups or colonies.

grid.   A specimen screen used in electron microscopy.

gross primary production. The rate (per unit area, per unit time) at which solar energy is photosynthetically converted into chemical energy.

ground meristem. A primary meristematic tissue (one of three) derived from the apical meristem; differentiates into the ground tissues, i.e. all tissues exclusive of epidermal and vascular. (*See* figs. M-2, P-15).

groundplasm. The cytoplasmic matrix of a cell which contains the organelles, membranes, etc.

ground tissue. Tissues other than vascular, epidermal and periderm. *Syn.* fundamental tissue.

ground tissue system. The entire complex of ground tissues. *Syn.* fundamental tissue system.

group selection. Selection operating at the group or species level and not at the individual level.

growth. Irreversible increase in number and size of cells due to division and enlargement.

growth curve. A curve which illustrates the change in the number of cells or organisms in a culture, as a function of time.

growth form. *See* life form.

growth layer, growth ring. *See* annual ring.

guard cells. Specialized crescent-shaped, chloroplast-containing cells that surround a stoma. Changes in the turgor of the guard cells result in the opening and closing of the stoma. (*See* figs. M-4, S-16).

guide cell. In mosses, an enlarged, elongated cell found in the center of the midrib.

gullet. A tubular groove in some algae (e.g. cryptomonads and euglenids) in which the flagella are inserted.

gum. A general term for a viscous polysaccharide secretion, often formed upon disintegration of a cell.

gummosis. The appearance of gums in plants as a result of physiological or pathological disturbances that induce a breakdown of cell walls and a discharge of their contents.

guttation. An exudation of liquid water through hydathodes from the tips of leaves. (*See* fig. G-4).

Fig. G-4. Guttation.

gymno-.   A prefix meaning naked.

gymnosperm.   A vascular plant which produces integumented seeds that are not enclosed by a megasporophyll or within carpel tissues; literally, a naked seed.

gymnostomous.   In mosses, a capsule that lacks peristome teeth.

gynandrous.   With stamens and pistils united, as in orchids (Orchidaceae).

gynecandrous.   With both staminate and pistillate flowers in the same spike, the former below the latter, as in sedges (Cyperaceae).

gynobase.   A dilation of prolongation of the receptacle of some flowers which bears the carpels or nutlets; as in many borages (Boraginaceae).

gynobasic style.   One that originates between the lobes of a deeply lobed ovary, as in mints (Lamiaceae) and borages (Boraginaceae).

gynodioecious, gynodioecy.   A species in which some plants bear perfect flowers whereas other plants bear only pistillate flowers.

gynoecium.   A collective term for all the carpels (female parts) in a flower. The carpels may be separate (apocarpous) or united (syncarpous).

gynogenesis.   Reproduction by parthenogenesis in which stimulation by sperm is necessary for development of an egg.

gynomonoecious, gynomonoecy.   A single plant which bears both perfect and pistillate flowers.

gynophore. A stalk or stipe bearing an ovary or fruit.

gynosporangium. *See* nucellus.

gynospore. *See* megaspore.

gynostemium. A compound structure in orchid flowers formed by the adnation of the stamens and pistil. *Syn.* column.

gypsophil. A plant which lives on chalky or limestone soils.

gyrate. Convolute.

gyrose. Having wavy, undulating lines.

H

habit. The general appearance or characteristic form of a plant, or other organism, e.g. erect, prostrate, climbing, etc.

habitat. The natural environment of an organism; the place where it is usually found.

hadrom. The trachery elements and the associated parenchymous cells of the xylem, exclusive of the supporting cells (fibers and sclereids). Contrast leptom.

hair. *See* trichome.

halberd-shaped. Hastate; sagittate (arrow-shaped), with divergent basal lobes.

half-life. The amount of time required for half of the radioactive molecules in a substance to decompose.

half-terete. Flat on one side and terete on other; semicircular in cross section.

halophyte. A plant which grows in salty (saline) or alkaline soils. Compare glycophyte.

halosere. A sere in which the pioneers invade saltwater or salt-flats.

hamate. Hooked at the tip.

hamulate, hamulous. Having small, hooked processes.

haplo-. A prefix meaning single or simple.

haplobiontic. Having one free-living organism in the life cycle; may either be haploid or diploid.

haplocheilic stoma. In gymnosperms, a stomatal type in which the subsidiary cells and the guard cells are unrelated ontogenetically.

haploid. Used in reference to either a nucleus, a cell or an entire organism in which only one member (the reduced *n* number) of each set of homologous chromosomes is present. In plants, the characteristic chromosome number of the gametophyte. *Syn*. monoploid. Compare diploid.

haploidization. In fungi, a process whereby a haploid nucleus is derived from a diploid nucleus by a gradual loss of chromosomes.

166

haplostele.   A cylindrical protostele with a smooth margin in transection.

haplostemonous.   With a single cycle (whorl) of stamens.

hapteron, *pl.* **haptera.**   A rootlike, branched, usually multicellular attachment organ found in some brown algae (Phaeophyta).

haptogamy.   Geitonogamy (self-fertilization) directly without a vector involved.

hardening.   An increase in the cold resistance of plant tissue.

hardpan.   Extremely hard soils through which plant roots, animals and water cannot penetrate. Often formed in flat land by the leaching of materials from upper soil layers into deeper layers; also formed from compaction of subsurface soil.

hardwood.   Wood (xylem) produced by angiosperms, especially deciduous, broad-leaved, dicot trees.

Hardy-Weinberg Law.   The mathematical expression of the relationship between relative frequencies of two or more alleles in a population; it demonstrates that the frequencies of dominant and recessive genes tend to remain constant in the absence of selective pressures.

harmonic biota.   A biota containing the basic adaptive types found in other comparable ecological areas. Compare disharmonic biota.

hastate.   Shaped like an arrowhead but with divergent lobes at the base.

hastula.   The terminal part of petiole located on the upper surface of a palmate-leaved palm; sometimes called a ligule.

haustorium, *pl.* **haustoria.**   **1.** A specialized root of parasitic vascular plants capable of penetrating and absorbing food and other materials from host tissues. (*See* fig. H-1). **2.** In fungi, an absorbing organ originating on a hyphae of a parasite, and penetrating into a cell of the host.

head.   **1.** A dense inflorescence of sessile or nearly sessile flowers on a compound receptacle. *Syn.* capitulum. **2.** The apical portion of a bacteriophage. (*See* fig. B-1).

heart-shaped.   Cordate; broadly ovate with two rounded lobes at the base.

heartwood.   The inner layers of xylem (wood) that are nonfunctional and nonliving; generally harder and darker colored than the surrounding, functional sapwood. (*See* fig. C-12).

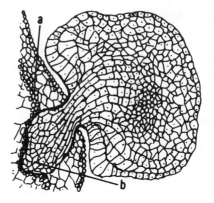

Fig. H-1. Haustorium: a. host tissue, b. haustorium.

hectare.   In the metric system, a measure of area equal to 10,000 square meters or about 2.5 acres.

heli-.   A prefix meaning sun.

helical thickening.   Secondary wall material deposited in the form of a helix or coil on the primary or secondary walls of tracheids or vessel elements. *Syn.* spiral cell wall thickening.

helicoid.   Curved or spiraled like a snail shell.

helicoid cyme.   A sympodial, determinate inflorescence in which the lateral branches develop on one side only, appearing simple; a bostryx; often incorrectly designated a scorpioid cyme.

helicospore.   A coiled or helical spore.

heliophyte.   A plant which grows in full sunlight.

heliotropic, heliotropism.   The turning of a plant, or plant organ in response to the sun; to follow the sun in a diurnal cycle.

helix.   A spiral moving around a cone or cylinder.

helophilous, helophytes.   Plants which live in fresh water marshes.

hematochrome.   Red pigments, probably related to xanthophyll, that occur in certain green algae (Chlorophyta) and euglenids.

hemi-.   A prefix meaning half.

hemicarp.   A half-carpel; a mericarp.

hemicellulose.   A polysaccharide resembling cellulose but more soluble and less complex; a common component of the cell wall matrix.

hemicryptophytes.   Perennial and biennial herbaceous plants having a periodic shoot reduction to a remnant shoot system, that lies relatively flat on the ground.

hemiparasite.   A parasitic plant containing chlorophyll and therefore partly self-sustaining, as in mistletoe (Viscaceae or Loranthaceae).

hemispheric.   Shaped like half a sphere.

hemizygous.   A gene present in a single copy in an organism, i.e. having no allele. *Syn.* haplozygous.

hepatic.   A liverwort. (*See* fig. H-2).

herb.   A nonwoody plant (annual, biennial or perennial), whose aerial portion naturally dies to the ground at the end of the growing season.

herbaceous.   A plant having the characteristics of a herb; having the texture of color of a foliage leaf.

herbage.   The vegetative or green parts of a plant.

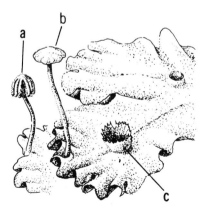

Fig. H-2. Hepatic: a. archegoniophore on female plant, b. antheridiophore on male plant, c. gemmae cup.

herbals. Early botanical works of the fifteenth, sixteenth and seventeenth centuries which described many of the plants known at the time, particularly medicinal and edible plants. Some contained attempts at classification.

herbarium. A collection of dried and pressed plant specimens, systematically arranged and labeled; used for taxonomic studies.

herbicides. Chemical compounds used to kill plants; they may be highly selective, injurious to only certain kinds of plants while not harming others.

herbivore, herbivorous. An organism which feeds only on plants.

hereditary. Transmission of a genetic trait from parent to offspring.

heredity. The genetic relationship between successive generations from the same parental stock.

heritability. The proportion of observed variability due to the cumulative effect of genes; the remainder being due to the environment.

herkogamy. The spatial separation of anthers and stigma in a perfect flower.

hermaphrodite. An organism or structure possessing both male and female reproductive organs; e.g. a flower with both stamens and pistil(s). *Syn.* bisexual, monoclinous, perfect.

hesperidium. A type of berry having a leathery pericarp, as in citrus fruits (Rutaceae).

heterandrous. Having two or more sets of stamens of different size and length.

hetero-. A prefix meaning other or different.

heteroblastic, heteroblasty. Having juvenile foliage distinctly different from mature (adult) foliage.

heterocarpous. Producing morphologically different types of fruits.

heterocellular ray. A ray in secondary vascular tissue composed of cells of more than one type. Compare homocellular ray.

heterocephalous. With staminate and pistillate flowers on separate heads on the same plant.

heterochromatin. Chromatin in whole chromosomes or parts of chromosomes which usually remains condensed prior to and during nuclear divisions. Most chromatin can alternate between the dispersed (euchromatin) and the condensed (heterochromatin) states. Compare euchromatin.

heterocyclic compound.   An organic compound, similar to an aromatic compound, which has at least one atom of an element other than carbon in the ring.

heterocyst.   A large, thick-walled, specialized cell produced by some filamentous blue-greens. Thought to serve as a point of fragmentation of the filament, as a spore capable of regenerating a new filament, and as an important site of nitrogen fixation.

heterodromous, heterodromy.   Having opposite spirals as in some tendrils or phyllotaxis.

heteroecious.   Requiring two living hosts to complete its life cycle, as in certain parasitic rust fungi.

heterogametangia.   Male and female gametangia which are morphologically distinguishable.

heterogametes.   Flagellated male and female gametes which are morphologically distinguishable.

heterogametic (plants).   Plants which produce gametes with differing chromosome complements, i.e. some gametes with an X and some with a Y chromosome.

heterogamous.   Bearing two or more kinds of flowers.

heterogamy.   Reproduction by morphologically distinguishable, flagellated gametes.

heterogeneous.   Not uniform; differing.

heterokaryosis, heterokaryotic.   In fungi, refers to a single mycelium having two or more genetically different nuclei in the same protoplast.

heterokont.   A biflagellated cell having flagella of unequal length.

heteromerous.   Having a different number of parts in different whorls.

heteromorphic.   **1.** Morphologically different; with different forms; polymorphic. *Opp.* monomorphic. **2.** A life history in which the haploid and diploid generations are dissimilar in form.

heterophyllous.   Having leaves of different sizes and/or shapes. (*See* fig. H-3).

heteropolar pollen.   A pollen grain or spore which has different polar faces; one face has an aperture, the other does not.

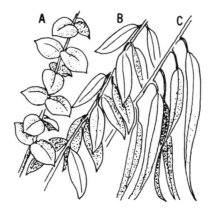

Fig. H-3. Heterophyllous: A. juvenile foliage, B. transitional foliage, C. adult foliage.

heterosis.   Hybrid vigor. The measurable superiority of an F_1 hybrid over either parent.

heterosporous, heterospory.   Having two types of haploid spores, usually referred to as microspores (male), which give rise to microgametophytes, and megaspores (female), which give rise to megagametophytes.

heterostylous, heterostyly.   A species which has flowers with styles differing in length or shape.

heterothallic, heterothallism.   A condition in certain algae and fungi in which male and female gametes are produced on different filaments or plants. Although these gametes are morphologically similar, sexual reproduction occurs only between two different sexual strains, usually designated as plus ($+$) and minus ($-$). Self-incompatible, self-sterile.

heterotrichous, heterotrichy.   A condition in some algae and bryophytes in which erect filaments arise from a prostrate portion.

heterotroph, heterotrophic.   An organism which cannot produce its own food and must consume other organisms or their organic products for food, e.g. animals, saprophytes and parasites.

heterozygote, heterozygous, heterozygosity.   A zygote or individual which possesses two different alleles at the same locus on homologous chromosomes; having unlike genes for a particular characteristic.

hexa-.   A prefix meaning six.

hexamerous.   With parts in sixes.

hexaploid.   Having six haploid chromosome sets.

hexavalent.   An association of six homologous or partially homologous chromosomes during meiosis.

hibernaculum, *pl.* **hibernacula.**   A winter bud; develops into a new plant under proper growing conditions; e.g. as in certain carnivorous plants.

hibernal, hiemal.   Appearing in the winter.

high-energy compounds.   Phosphates which have chemical bonds containing greater than normal potential energy; e.g. ATP and ADP.

hilum.   1. The central part of a starch grain surrounded by more or less concentric layers. **2.** A scar on a seed marking the point of attachment to the funiculus.

hip.   A rose fruit formed by a group of achenes surrounded by a receptacle and hypanthium. Technically, a cynarrhodium.

hippocrepiform.   Horseshoe-shaped.

hirsute.   Covered with rather rough and stiff trichomes.

hirsutulous, hirtellous.   Minutely hirsute.

hispid.   Covered with long, rather stiff trichomes; usually stiff enough to penetrate the skin.

hispidulous.   Somewhat or minutely hispid.

histogen.   Meristematic tissue in shoot or root tips that forms a definite tissue system in the plant body.

histogenesis.   The formation of tissues.

histogram.   A graph in which classes of events or observations are indicated along a horizontal axis, with the frequency of each class represented by rectangles of proportionate height. *Syn.* frequency histogram.

histioid.   *See* arachnoid.

histology.   The study of tissues.

histone.   A protein associated with DNA in chromosomes.

hoary.   Covered with a white or grayish-white pubescence. *Syn.* canescent.

holdfast.   1. The basal part of an algal or fungal thallus that attaches it to a solid object. (*See* fig. H-4). **2.** Cuplike structures at the tips of some tendrils; used for attachment.

holobasidium, *pl.* **holobasidia.**   A simple, club-shaped structure in which karyogamy and meiosis occur, and which bears basidiospores on its surface.

holocarpic.   A condition in which the entire thallus develops into a reproductive structure, as in certain unicellular fungi.

hologamodeme.   A group of individuals of a taxon which are believed to freely interbreed with specified conditions. This term is preferred by some to 'biological species' which relates to a taxonomic category. *Syn.* ecospecies.

holoparasite.   A plant lacking chlorophyll and entirely dependent on its host for nourishment. They are subdivided into stem or root parasites depending on the portion of the host they attack.

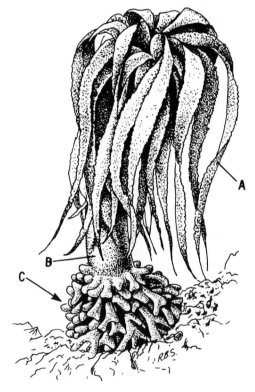

Fig. H-4. Holdfast: A. blade, B. stipe, C. holdfast.

holophyte.   A plant in which the entire, above-ground portion is dispersed.

holosericeous.   Covered with fine and silky pubescence.

holotype.   The single specimen, the type specimen, designated to carry the name of a new species.

holozoic.   Ingesting solid food particles, or entire organisms.

homandrous.   Having stamens all the same size and shape.

homeo-, homo-, homolo-.   Prefix meaning similar or same.

homeohydric.   An organism which can regulate its cell water content, regardless of changes in atmospheric humidity.

homeostasis.   Maintenance of a relatively stable internal physiological environment or equilibrium, in an organism, population or ecosystem.

homocellular ray.   A ray in secondary vascular tissue composed of cells of a single form. Compare heterocellular ray.

homochlamydeous.   Having a perianth composed of similar parts, each called a tepal.

homogametic.   The sex which produces only one type of gamete. Compare heterogametic.

homogamous, homogamy.   Simultaneous maturation of stamens and stigma in a bisexual flower.

homogeneous.   Uniform, alike; having the same nature or consistancy.

homoiomerous.   Having algal cells scattered throughout a lichen thallus.

homokaryotic.   In fungi, a condition in which all nuclei are genetically alike in a mycelium.

homologous, homology.   Having the same phylogenetic, or evolutionary origin, but not necessarily the same structure and/or function.

homologous chromosomes.   Morphologically similar chromosomes which associate in pairs during prophase I of meiosis; each member of a pair is derived from a different parent.

homomallous.   In mosses, having all leaves pointing in the same direction.

homomorphic. With a single form; monomorphic. *Opp.* heteromorphic.

homonym. In nomenclature, a name rejected because it duplicates a name previously and validly published for a group of the same taxonomic rank and based on a different type.

homoploid series. A group of related species which all have the same chromosome number, e.g. *Pinus* (pines).

homosporous, homospory. Having only one type of spore. *Syn.* isosporous.

homostylous. Having styles of the same lengths and/or shapes.

homothallic, homothallism. A condition found in certain algae and fungi in which compatible gametes are produced on a single thallus; single individuals can form fertile zygotes, and are thus self-fertile.

homozygote, homozygous, homozygosity. A zygote or individual which possesses identical genes at the same locus on homologous chromosomes. An organism may be homozygous for one or several genes, or rarely for all genes.

hood. 1. A concave segment of the corona of a milkweed (Asclepiadaceae) flower. *Syn.* crest. 2. An appendage which hangs over the opening in a pitcher plant trap. *Syn.* lid.

hooked. With an incurved apex, in reference to trichomes.

horizontal system. *See* radial system.

hormogonium, *pl.* **hormogonia.** A multicellular segment of some filamentous blue-greens, capable of developing into a new filament.

hormone. An organic chemical normally produced in minute amounts in one part of an organism, and transported to another area of the same organism where it affects growth and/or other functions.

horn. A curved, pointed and hollow projection which is part of the corona of a milkweed (Asclepiadaceae) flower.

horticulture. The branch of agriculture concerned with garden crops, orchard plants and ornamental plants, which are grown for food, medicinal or aesthetic purposes.

host. A plant which nourishes a parasite.

hourglass cell. *See* pillar cell.

humidity. The amount of moisture in the air expressed as the percent of the total amount possible at that same temperature.

humification. The processes involved in decomposition of organic materials in the soil that result in the production of humus.

humus. Decomposing organic matter in the soil.

husk. The outer covering of certain fruits or seeds usually derived from the perianth or involucre.

hyaline. Thin and translucent or transparent, like glass.

hyaloplasm. *See* cytoplasm.

hybrid. **1.** An individual resulting from the union of gametes differing in one or more genes; a heterozygous individual. **2.** The offspring of a cross between two different species, races or varieties; a crossbred plant.

hybridization. The process of crossing individuals with different genetic makeup.

hybrid segregate. A variation arising in the second or subsequent generation after hybridization.

hybrid swarm. A generally variable population typically occurring in the overlapping geographical area between interfertile species or subspecies. Made up of later generation hybrids and/or backcrosses.

hybrid vigor. *See* heterosis.

hydathode. A minute pore or specialized structure through which water (containing various dissolved salts, sugars and other organic substances) is discharged from a leaf by guttation; may be differentiated as secretory trichomes.

hydrocarbon. An organic compound composed of hydrogen and carbon.

hydrocarpic. A condition of aquatic plants in which pollination occurs above the water surface with fruit development occurring underwater.

hydrochore. A plant disseminated by water.

hydrogen bond. A weak bond between two atoms or molecules linked together by a hydrogen atom. Important in water, proteins and chromosomes.

hydroids. The central water conducting cells in stems of many mosses (Bryophyta).

hydrolysis.   The splitting of a molecule into smaller molecules by the addition of hydrogen (H^+) or hydroxyl (OH^-) ions derived from water.

hydromorphic.   Referring to the structural features of hydrophytes. *Syn.* hygromorphic.

hydrophilic.   Water loving. Refers to materials such as molecules that readily associate with water.

hydrophilous, hydrophily.   **1.** Growing in moist places or in water. **2.** Pollination by water.

hydrophobic.   Water repelling. Refers to materials such as molecules that are only slightly, if at all soluble in water.

hydrophyte.   A plant living in water or in a very moist habitat; an aquatic plant. (*See* fig. H-5).

hydroponics.   The cultivation of plants in aqueous nutrient solutions containing all essential elements, rather than in soil.

hydrosere.   A sere in which the pioneers invade water.

hydrotropism.   Growth towards, or in response to the influence of water.

hygromorphic.   *See* hydromorphic.

hygroscopic.   Readily absorbing and retaining moisture from the atmosphere; results in changes in form or position of certain cells or structures.

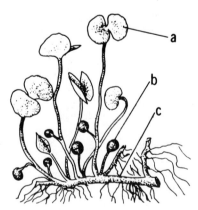

Fig. H-5. Hydrophyte: a. leaf, b. sporocarp, c. rhizome.

hygroscopic water.   Water that adheres so tightly to soil particles that it is seldom available to plants.

hylophilous, hylophyte.   Plants which dwell in forests or wooded areas.

hymenium, *pl.* **hymenia.**   The spore-bearing layer of asci on an ascocarp (Ascomycetes), or of basidia on a basidiocarp (Basidiomycetes), besides any associated sterile hyphae.

hymenopterophily.   Pollination by bees.

hypanthium.   A floral tube formed by the fusion of the basal portions of the sepals, petals and stamens, and from which the rest of the floral parts emanate.

hyper-.   A prefix meaning above or over.

hyperparasite.   A plant that parasitizes other parasites, as some mistletoes (Viscaceae or Loranthaceae). *Syn.* epiparasite.

hyperplasia, hyperplastic.   An excessive, abnormal, usually pathological, multiplication of the cells of a tissue or organ.

hyperspace.   Any space defined by three or more dimensions.

hypertonic.   A solution having an osmotic concentration high enough to gain water across a membrane from another solution; most living cells placed in such a solution will lose water (plasmolyze).

hypertrophy.   An excessive, abnormal, usually pathological, enlargement of cell size in a tissue or organ.

hypha, *pl.* **hyphae.**   A single, threadlike filament which is the structural unit of fungi; the hyphae together comprise the mycelium.

hyphodromous.   A type of leaf venation with a single primary vein and all other venation absent, rudimentary or concealed within a coriaceous or fleshy blade.

hyphopodium, *pl.* **hyphopodia.**   A small appendage on hyphae of certain fungi.

hypnospore.   A spore formed within a parental cell.

hypo-.   A prefix meaning under or less.

hypocotyl.   That portion of an embryo or seedling below the cotyledon(s) and above the radicle (but sometimes including it); the embryonic stem in a seed.

hypocrateriform.   See salverform.

hypodermis.   A layer or layers of cells beneath the epidermis, which are morphologically distinct from the underlying cortical or mesophyll cells. (*See* fig. H-6).

hypogeal, hypogean.   Describing seed germination in which the cotyledons remain beneath the surface of the soil. Compare epigeal.

hypogeous.   Growing or developing below the soil surface.

hypogynium.   A perianth-like structure of bony scales subtending the ovary as in some sedges (Cyperaceae).

hypogynous, hypogyny.   Having the flower parts attached below the base of the ovary and free from it; flowers with this arrangement have a superior ovary. (*See* fig. E-4).

hypomorphic gene.   A mutant gene which acts in the same direction as a normal allele, but it has less effect. *Syn.* leaky gene.

hyponasty.   The upward movement or bending of leaves and petals which results from unequal and more rapid growth of the lower surface. Compare epinasty.

hyponym.   In taxonomy, a name not based on a type species and/or, a name rejected for lack of an identified type.

hypophloeodal.   A thallus that grows beneath bark; especially in reference to fungi.

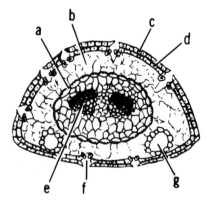

Fig. H-6. Hypodermis: a. endodermis, b. mesophyll, c. epidermis, d. hypodermis, e. vascular bundle, f. stoma, g. resin duct.

hypophysis.   **1.** The uppermost cell of a suspensor from which part of the root and rootcap are derived in the embryo of angiosperms. **2.** In mosses, a swelling of the seta under the capsule.

hypostase.   A small mass of specialized nucellar tissues located at the chalaza of certain ovules, which by growth and intrusion, causes ovule curvature or distortion.

hypostomatic stoma.   Referring to a leaf in which stomata occur only on the lower surface of the leaf.

hypothallus, *pl.* **hypothalli.**   **1.** A thin, often transparent deposit at the base of the fructifications in some slime molds (Myxomycetes). **2.** The dark outer thalloid margin of some crustose lichens.

hypotheca.   The smaller half of the frustule of diatoms.

hypothecium.   A thin layer of hyphal tissue immediately beneath the hymenium of an apothecium. *Syn.* hypothecial layer.

hypothesis.   A working explanation or supposition based upon accumulated facts and suggesting some general principle or relation of cause and effect; a postulated solution to a scientific problem that must be tested by experimentation and, if not supported, discarded.

hypotonic.   A solution having an osmotic concentration low enough to lose water across a membrane to another solution.

hypotropous.   An erect ovule, with the micropyle below, and the raphe dorsal or ventral.

hypsophylls.   Leaves borne at the upper levels of a plant, as with various floral bracts; some possibly protective in function. *See* cataphylls.

hysteranthous.   With leaves appearing after the flowers.

hysterothecium.   An oblong or linear perithecium which opens by a cleft. Sometimes considered an apothecium.

I

IAA.   *See* indoleacetic acid.

idioblast.   A unique cell marked by differences in size, form or contents from other cells in the same tissue.

idiogram.   A diagram illustrating the general morphology of the chromosome complement of a cell including overall size, position of centromeres, satellites and secondary constructions. *Syn.* karyogram.

im-.   A prefix meaning not without, in, into.

imbibition.   The absorption of water by dry or partially dry colloidal materials by the adsorption of water molecules onto and into the materials. Cell walls and protoplasm absorb water by imbibition causing them to become softer, swollen and more elastic. Compare active absorption.

imbricate.   With margins of structures overlapping like shingles on a roof. Compare valvate. (*See* fig. I-1).

immersed.   **1.** Completely submerged in water. **2.** In mosses, capsules having short seta and the tips of the perichaetial leaves extending beyond the lid.

immune.   Exempt from disease.

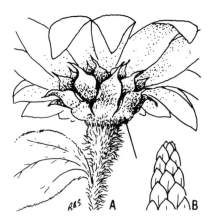

Fig. I-1. Imbricate: A. phyllaries in a composite, B. bracts on an inflorescence.

imparipinnately compound.   Odd-pinnately compound with a terminal leaflet.

imperfect flower.   A flower which lacks either stamens or carpels.

imperfect fungi.   *See* deuteromycete.

imperfect stage.   The part of life cycle which reproduces only by asexual means.

imperforate.   Lacking perforations.

implexed, implicated.   Having leaf margins folded sharply inward.

importance value (I.V.).   As used in vegetational analysis, it is the sum of relative density, relative frequency and relative dominance.

impressions (fossil).   A fossil with little or no organic matter preserved; only an outline and the vein impressions remain visible.

in-.   A prefix meaning not without, in, into, on.

inaperturate (pollen).   Pollen grains which lack germinal pores.

inbred line.   A nearly homozygous plant or group of plants produced by continued inbreeding.

inbreeding.   In plants, a breeding system in which sexual reproduction involves the interbreeding of closely related plants by self-pollination and backcrossing.

inbreeding coefficient.   A quantitative measure of the degree of inbreeding.

incanescent.   Becoming gray or canescent.

incised.   Leaf margins cut sharply, irregularly and rather deeply.

inclinate.   With a leaf blade folded or curved transversely near the apex.

included.   Not protruding beyond the surrounding structure. Compare exserted.

included phloem.   Secondary phloem found within the secondary xylem of certain dicots. Also called interxylary phloem.

incompatibility.   A condition, controlled by a variety of factors, in which some otherwise normal pollen grains are unable to function properly on certain pistils.

incomplete dominance.   The condition that results when two different alleles acting together produce an effect intermediate between the effects of these same genes in the homozygous condition.

incomplete flower.   A flower that lacks one or more of the following: sepals, petals, stamens or pistils.

incrassate.   Thickened.

increment.   An addition, as in growth of a plant body.

incubous.   Having leaves inserted so that the upper part or margin of a leaf covers the base of the leaf directly above it; as in liverworts (Jungermanniales). Compare succubous.

incumbent.   1. Lying or leaning upon. **2.** With anthers turned inward. **3.** Having cotyledons where the back of one rests against the radicle.

incurved.   Curved inward or upward.

indefinite.   Indeterminate or uncertain in regard to number.

indehiscent.   Not dehiscent, remaining closed at maturity.

independent assortment.   *See* Mendel's principle of independent assortment.

indeterminate growth.   Unrestricted growth, as with a vegetative apical meristem capable of producing an unlimited number of lateral organs. Contrast determinate growth.

indeterminate inflorescence.   One whose lateral flowers open first while the primary axis continues to grow, hence the terminal flower is the last to open. Contrast determinate inflorescence.

index fossil.   A fossil which identifies the geological stratum in which it is found. Sometimes called an indicator species.

indicator.   A plant which is characteristic of a particular ecological seral stage.

indigen, indigenous.   Native to a region or country; not introduced or imported. *Syn.* native. Compare endemic.

indoleacetic acid (IAA).   A naturally occurring growth hormone of the auxin type. *See* auxin and indolebutyric acid.

indolebutyric acid (IBA).   A synthetic auxin widely used in horticulture to induce rooting of cuttings.

inducer. A substance that derepresses an operator gene by inactivating the repressor produced by a regulator gene. (*See* fig. O-2).

induction. 1. The stimulation of a lysogenic bacterium to produce an infective phage. 2. Any stimulus that produces a particular effect.

indument, indumentum. Any covering of a plant surface, especially pubescence.

induplicate. With margins rolled or folded inward so that the margins touch.

indurate. Hardened.

indusium, *pl.* **indusia.** An epidermal outgrowth which covers a sorus in some ferns.

induviae. A persistent part of the perianth or leaves which withers but does not fall off the pedicel or petiole.

inequilateral. Asymmetrical, unequal-sided.

inermous. Unarmed, without spines or prickles.

inferior ovary. An ovary that is situated below the point of insertion of the other floral organs; one that is adnate to the hypanthium and situated below the calyx-lobes. (*See* figs. C-2, E-4).

inflated. Puffed out, bladderlike.

inflexed. Bent abruptly inward.

inflophyte. A plant in which the old inflorescence is the portion normally dispersed.

inflorescence. The arrangement of flowers on a floral axis; a floral cluster.

infra-. A prefix meaning below or beneath.

infraspecific. Referring to any taxon below the species level.

infundibular. Funnel-shaped.

inhibitor genes. Genes which suppress the action and expression of other genes, and are thus a type of controlling gene. *Syn.* suppressor genes.

initial. 1. A cell in a meristem which mitotically divides to produce two cells, one of which remains in the meristem, with the other added to the plant body. 2. Sometimes refers to a cell in its earliest stage of specialization; e.g. a meristem cell which differentiates into a mature specialized element.

innate. Borne at the tip of a supporting structure, as in some anthers at the apex of a filament.

inner bark. The portion of living bark inside the innermost periderm. (*See* fig. C-12).

inner membrane sphere. *See* oxysomes.

inner veil. In some Basidiomycetes, the hyphal tissue joining the margin of the cap to the stalk or stipe. The remains of the membrane left on the stalk after the expansion of the cap are called the annulus or ring. Also called the partial veil.

innocuous. Unarmed or spineless.

innovation. A new vegetative shoot, as in some perennial grasses.

inoculation. The introduction of bacteria or other microorganisms onto seeds, into plants, into soil or onto a culture medium.

inoculum. A pathogen or its part, which can infect a plant.

inoperculate. In mosses, lacking an operculum; without a lid.

inorganic compound. A chemical compound which is usually derived from abiotic processes; e.g. salts and noncarbon compounds.

insecticide. Any substance used for the control or eradication of insects.

insectivorous. Referring to plants that capture insects. (*See* fig. I-2).

insect pollination. Pollination in which the vectors of pollen transfer are insects.

inserted. Growing upon or attached to.

in situ. In the normal or natural position.

insular. Pertaining to an island.

intectate pollen. Pollen grains without a tectum or tectal areas.

integument. The outer cell layer (or layers) which surrounds the nucellus of the angiosperm ovule and develops into the seed coat. (*See* figs. A-8, E-2, O-3).

integumentary tapetum. A nutritive layer that develops on the inner epidermis of the

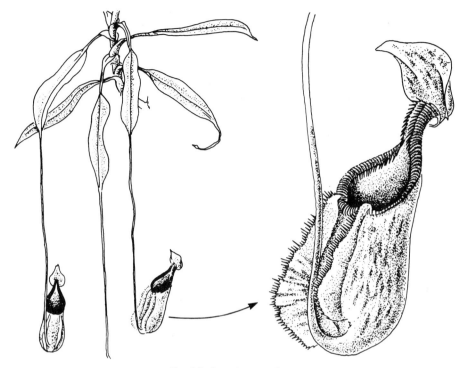

Fig. I-2. Insectivorous plant.

integuments of certain plants (e.g. the Sympetalae); it is formed when the nucellus disintegrates, thereby bringing the embryo sac into contact with the integuments.

inter-. Prefix meaning between or in between.

intercalary, intercalated. Lying between two cells, tissues or other structures.

intercalary growth. Growth resulting from the action of intercalary meristems; growth that occurs some distance from an apical meristem, or between the base and apex of an organism. This is the typical growth pattern of grasses. (*See* fig. I-3).

intercalary meristem. A localized meristematic region in an elongating internode; a meristem that produces cells both above and below it, or one that lies between tissues that are more or less mature (differentiated). (*See* fig. I-3).

intercellular canal. A tubular intercellular space of indeterminate length into which various resins and gums are secreted.

intercellular layer. *See* middle lamella.

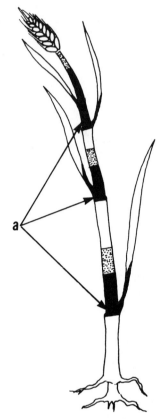

Fig. I-3. Intercalary growth: a. intercalary meristem, white area—oldest growth, black area—active new growth, shaded area—slightly older growth.

intercellular space.   A space between two or more adjacent cell walls in a tissue system; they are most commonly formed during the process of cell differentiation.

intercellular substance.   *See* middle lamella.

intercostal.   Space between ribs or costae.

interfascicular cambium.   A cambium developing between vascular bundles from interfascicular parenchyma cells.

interfascicular region.   A tissue region located between vascular bundles in a stem; connects the cortex and pith in primary stems. *Syn.* medullary or pith ray.

interference.   The effect one crossover has on the probability that another will occur in the immediate vicinity.

interfoliar.   On the stem between the leaves; among the leaves.

intergeneric hybrids.   Hybrids between species of two or more genera.

interkinesis.   The stage between the two meiotic divisions.

intermediary cell.   A parenchyma cell found in the phloem of smaller vascular strands. Serves as a connection between photosynthetic tissue and the sieve tube system.

internal phloem.   Primary phloem located inside (i.e. toward the center of the stem) the primary xylem. Replaces the term intraxylary phloem.

internerves.   Spaces between nerves.

internode.   The portion of the stem between two successive nodes.

interphase.   The period between two active mitotic or meiotic divisions.

interpositional growth.   *See* intrusive growth.

interrupted.   Not continuous or regular.

interruptedly pinnately compound.   Pinnately compound leaves with small leaflets alternating between larger ones along the rachis.

interspecific.   Referring to some relationship between members of different populations, as in interspecific competition or interspecific hybridization.

interstades.   A glacial period in which ice is retreating.

interstitial (growth).   A type of growth occurring in organs that do not have specific or localized meristematic regions; all-over growth, as in some fruits.

intertidal region.   That part of the sea floor or coast line exposed between the highest and lowest tide levels. (*See* fig. A-16).

intervascular pitting.   The pitting between tracheary elements.

interxylary.   Within and surrounded by xylem tissue.

interxylary cork.   Cork that develops within the xylem tissue.

interxylary phloem.   *See* included phloem.

intine.   The inner wall layer of a pollen grain or spore. (*See* fig. S-6).

intra-.   A prefix meaning within.

intranuclear division.   Nuclear division occurring within the nuclear envelope.

intraspecific.   Referring to some relationship between the members of the same population or species, as in intraspecific competition; refers to any taxon within a species.

intraxylary phloem.   *See* internal phloem.

introduced.   Brought into one area or region, from another.

introgressive hybridization.   Genetic modification of one species by another through the intermediacy of hybrids; introgression.

introrse.   Turned inward toward the axis; as the opening of an anther toward the inside of a flower.

intrusive growth.   Growth of one cell between the cell walls of other adjacent cells. Also called interpositional growth. (*See* fig. I-4).

intumescence.   A knoblike or pustulelike outgrowth on plants caused by environmental disturbances.

intussusception (of cell wall).   Cell wall growth by interpolation of new cell wall material inside previously formed cell walls. Compare apposition.

invaginated.   Enclosed within a sheath.

invagination.   An ingrowth.

inversion.   A rearrangement of a chromosome segment so that its genes are in reversed order in relation to the rest of the chromosome.

inverted.   Upside down; turned over.

in vitro.   In glass; with reference to experiments done in glass vessels, such as test tubes or petri dishes.

in vivo.   Within a living organism.

involucel.   A small involucre; a secondary involucre that subtends a part of an inflores-

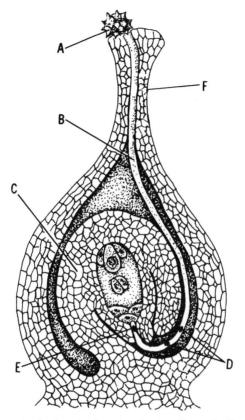

Fig. I-4. Intrusive growth of polen tube through style: A. pollen grain, B. pollen tube, C. ovule, D. sperm, E. tube nucleus, F. style.

cence, as the bracts subtending the secondary umbels in the umbel family (Apiaceae or Araliaceae).

involucral.   Pertaining to an involucre.

involucrate.   Having an involucre.

involucre.   **1.** One or more whorls of small bracts (phyllaries) which subtend a flower or inflorescence. **2.** A cup-shaped disc which subtends the ovule in the Gnetophyta. **3.** The tissue that encloses the archegonia in liverworts.

involute.   Margins rolled or turned in over the upper or ventral surface. *Opp.* revolute.

ion.   An atom or group of atoms with a negative or positive charge.

ionizing radiation.   Very high-energy radiations that can remove electrons from atoms and attach them to other atoms, thereby producing positive and negative ion pairs. Ionization is probably the chief cause of injury to protoplasm.

irregular.   Asymmetrical, showing a lack of uniformity.

irregular flower.   A bilaterally symmetrical flower in which the floral parts in a given whorl are dissimilar in size and/or shape. *Syn.* zygomorphic.

irrigation.   The act of supplying water to a crop.

irritability.   The ability of living protoplasm to respond to external stimuli.

isidium, *pl.* **isidia.**   A rigid protuberance which may separate and serve as a means of asexual reproduction. Found on the upper portion of a lichen thallus.

iso-.   A prefix meaning equal.

isoallelles.   Alleles indistinguishable except by special tests.

isobilateral leaf.   A leaf in which palisade parenchyma occurs on both sides of the blade. *Syn.* isolateral leaf.

isodiametric.   With equal diameters, as a cell where length, width and height are equal.

isodynamous.   Having equally developed structures.

isoenzymes.   *See* isozymes.

isogametangium, *pl.* **isogametangia.**   Gametangia, which are assumed to be of opposite sexes, but are morphologically indistinguishable.

isogamete.   Gametes which are morphologically indistinguishable and usually flagellated. If motile, they may also be called isoplanogametes.

isogamous, isogamy.   A type of sexual reproduction in algae and fungi in which the gametes (or gametangia) are similar in size and morphologically indistinguishable. Compare anisogamous and oogamous.

isogenic lines.   Very closely related genetic lines differing only at a single locus.

isokont.   Mobile cells having flagella of equal length.

isolateral leaf.   *See* isobilateral leaf.

isolation. The separation of one population from another so that interbreeding is prevented; isolation may result from one or more factors such as geographical isolation, physiological isolation, or behavioral isolation.

isomer. One of two or more chemical compounds which have the same molecular formula, but have a different molecular structure; e.g. glucose and fructose.

isomerous. Having the same number of parts in each whorl, in reference to flowers.

isomorphic. **1.** Morphologically indistinguishable. **2.** A life history in which the haploid and diploid generations are similar in form.

isophasic growth. Integrated host-parasite growth.

isophyllous. Producing one type of leaf, of the same size and shape.

isoplanogamete. A motile isogamete.

isopolar pollen. Pollen in which the equatorial plane divides a grain into identical halves so that the proximal and distal faces are similar.

isositchous. Having equal rows.

isosporous. *See* homosporous.

isotonic. Having the same osmotic concentration.

isotope. One of several possible forms of a chemical element having the same chemical properties, but differing in atomic weight.

isotype. A specimen of the type collection other than the holotype; a duplicate of the holotype.

isozymes. Different forms of the same enzyme. *Syn.* isoenzymes.

isthmus. The constricted portion connecting the two semicells in many desmids.

J

jacket cell, jacket layer. A layer of cells surrounding a reproductive structure; e.g. as those around the antheridium in some lower vascular plants, or around the egg (archegonium), in some gymnosperms.

jaculiferous. Having dartlike spines.

joint. An articulation; the node of a grass culm.

jointed. Having nodes, or points of articulation.

jordanon. *See* microspecies.

jugate. Having two leaflets originating from a common point.

julaceons. In mosses, a smooth cylindrical shoot resulting from closely and evenly imbricated leaves.

juxtaposition. Being placed or occurring side by side.

K

karyallagy. The fusion of sexually undifferentiated cells, as swarm cells in slime molds.

karyogamy. The fusion of two nuclei following plasmogamy.

karyogram. *See* idiogram.

karyokinesis. The process of nuclear division as distinguished from cell division (cytokinesis). *Syn.* mitosis.

karyolymph. The nuclear sap.

karyotype. The general appearance of the chromosome complement of an individual or a group of related individuals, with regard to their number, size, shape, etc.; usually based on observations of chromosomes in mitotic metaphase.

kcal. A kilocalorie.

keel. **1.** A sharp crease or ridge, like the bottom of a boat. **2.** The two united petals of a papilionaceous flower in legumes. **3.** A ridge on the trap of a pitcher plant.

keeled. Having a keel, sharply creased.

kelp. A common name for any of the large marine brown algae (Phaeophyta).

kernel. **1.** An entire seed of a grass, as a kernel of corn. **2.** The inner and usually more edible portion of a seed contained within the seed coat.

key. **1.** A series of paired, contrasting statements used to facilitate the identification of plants; also referred to as dichotomous keys. **2.** A samara.

kilo-. A prefix meaning one thousand.

kilocalorie (kcal). The amount of energy needed to raise 1000 grams of water one degree Celsius.

kinetochore. *See* centromere.

kingdom. The highest or most inclusive taxonomic category of plants and animals.

Fig. K-1. Krummholz.

kinin.   *See* cytokinin.

knee.   **1.** An abrupt bend in a stem or tree trunk. **2.** An outgrowth found above ground level on some tree roots.

knot.   The basal portion of a dead branch, which was embedded in the wood of a tree by the successive deposition of annual layers of xylem.

Krebs cycle.   The breakdown of pyruvic acid under aerobic conditions producing carbon dioxide, water and energy in the form of ATP. *Syn.* citric acid cycle and tricarboxylic acid cycle.

krummholz.   The creeping, often stunted growth habit found in certain high altitude tree species near their upper limit of distribution. Caused mainly from the effects of wind. (*See* fig. K-1).

K selection.   A generalized type of reproductive strategy found in species which belong to stable and saturated communities. Characterized by low mortality, low fecundity, slow development and greater competitive ability. Compare r selection.

L

labeled compound. A compound containing a radioactive isotope; used to follow a compound through a series of biological changes.

labellum. In orchid flowers, the lower of the three petals which usually is greatly modified and enlarged. It serves as an attraction mechanism and landing platform for insects. *See* lip. (*See* fig. L-1).

labiate. Lipped; having a lipped structure, as where a corolla or calyx is divided into two differently shaped parts forming an upper and lower lip.

labium. *pl.* **labia.** A lip.

laccate. With a varnished appearance.

lacerate. Cut irregularly, appearing torn, as in certain leaves and ligules.

laciniate. Cut deeply into closely parallel, narrow divisions, as in leaves.

lactiferous. **1.** Having a milky sap or juice. **2.** Bearing latex.

lacuna, *pl.* **lacunae.** **1.** An internal air space or chamber, usually between cells as in leaf, stem, and root tissues. Also, any space within a cell. **2.** Leaf gaps as viewed in cross sections of stem nodes.

lagging. A delay in the movement of a chromosome from the equator to the poles during anaphase. Such chromosomes are excluded from the daughter nuclei.

lambda (λ). A microliter ($\mu\ell$). A microliter of water weighs one milligram. The volume contained in a cube one millimeter on a side.

lamella, *pl.* **lamellae.** **1.** A thin plate or layer. **2.** Layers of photosynthetic membranes in chloroplasts containing chlorophyll and other pigments. **3.** The gills of mushrooms (Basidiomycetes).

lamellate, lamellose. Having or made up of thin plates.

lamina. *See* blade.

laminar. Thin and flattened, as in a leaf blade.

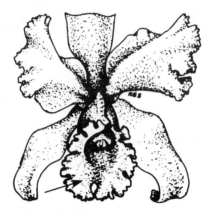

Fig. L-1. Labellum.

laminarin.   The primary food storage product of the brown algae; a polysaccharide composed of glucose and mannitol.

lanate, lanose.   Woolly, cottony; covered with long, fine, intertwined trichomes (hairs).

lanceolate.   Lance-shaped; much longer than broad; widened above the base and tapering toward the apex.

lanuginous.   Cottony, similar to lanate but with shorter trichomes (hairs).

lanulose.   Very short-cottony.

lasio-.   A prefix meaning woolly or hairy; hence, lasiocarpus, with pubescent fruit; lasiolepis, with woolly scales.

lateral.   Borne on, or pertaining to, the side of an organ or structure; e.g. a lateral bud.

lateral bud.   A bud in the axil of a leaf. *Syn.* axillary bud. (*See* figs. B-8, T-8).

lateral conjugation.   The conjugation of adjacent cells in algal filaments by the formation of lateral protuberances, as in certain *Spirogyra* species.

lateral meristems.   Meristems that give rise to secondary tissue; e.g. the vascular cambium and cork cambium.

lateritic soil.   A relatively infertile soil characteristic of tropical and subtropical areas, usually red in color due to the presence of iron oxides. They are highly acidic and have little organic matter.

late wood.   The wood in a perennial plant produced later in a growing season. It is denser and has smaller cells than early wood. Replaces the term summer wood.

latex.   A usually milky-looking liquid (although it may also be clear, yellow, orange or brown) contained in laticifers. Latex consists of numerous organic and inorganic substances in solution and colloidal suspension. Rubber is an important component of the latex of more than 2000 species of angiosperms.

lathhouse.   An outdoor growing structure for plants and cuttings which by providing shade, lowers air temperatures thereby reducing transpiration and soil evaporation. Construction is with lath (wood or aluminum) or shade cloth (polypropylene fabric). *Syn.* shade house.

lati-.   A prefix meaning broad.

laticifer.   A cell or series of fused cells containing latex. They occur in various tissues and organs of plants and are restricted to the phloem of some species. They intergrade in morphology with certain idioblasts which do not contain latex. (*See* fig. L-2). *See* articulated and nonarticulated laticifer.

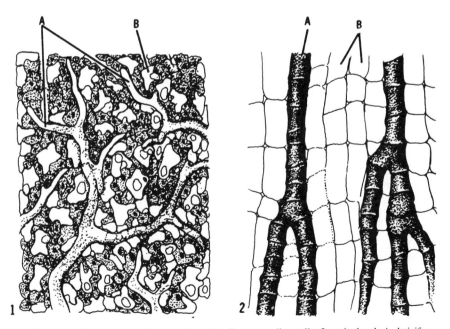

Fig. L-2. Laticifer: 1. nonarticulated: A. laticifer, B. surrounding cells; 2. articulated: A. laticifer, B. surrounding cells.

Laurasia. The northern group of continents which included what is now known as North America and Asia, and were part of the supercontinent of Pangaea.

lax. Loose, not densely packed. Contrast congested.

layering. A process of propagating plants from stem sections while they are still attached to the parent plant. Adventitious roots are induced to form after the stem is covered with soil or organic matter. Layering is frequently a natural process, and in horticulture it is of some importance in propagating blackberries, raspberries, currents, etc. *Syn.* layerage.

leach, leaching. **1.** The application of large amounts of water to soil to remove excess salts or fertilizers. **2.** The natural downward movement of minerals through the soil by percolating water.

leader. The main growing stem of a shrub or tree. Because of its apical dominance, it exerts control over the growth of lateral buds and branches beneath it.

leaf, *pl.* **leaves.** The principle photosynthetic (food manufacturing) and transpiring organ of a green plant. Although shapes vary, they are usually green, more or less flattened, expanded structures formed at a node as a lateral outgrowth of a stem. They commonly consist of a blade and petiole. Other functions may include water and food storage, protection, attachment, and asexual reproduction.

leaf axil. The upper angle between a leaf or leaf petiole and the stem to which it is attached. (*See* fig. L-3).

leaf buds. A bud that produces only leaves.

leaf buttress. A lateral protrusion below the apical meristem, which represents the initial stage in the development of a leaf primordium.

leaf fibers. The technical designation for economic fibers derived principally from the leaves of monocotyledons. Examples include Manila and bowstring hemp, sisal and henequen.

leaf gap. In the nodal region of a stem, a break or gap in the continuity of the primary vascular cylinder above the level where a leaf trace diverges toward a leaf. This gap is filled with parenchyma tissue. *Syn.* lacuna. (*See* figs. M-10, P-15).

leaflet. One of the distinct and separate divisions of a compound leaf.

leaf primordium. A lateral outgrowth from the apical meristem that develops into a leaf. (*See* fig. P-15).

leaf scar. A scar or mark left on a stem when a leaf falls, indicating the former place of attachment of the petiole or leaf base. (*See* fig. B-8).

leaf sheath. The basal portion of a leaf blade or petiole that more or less completely surrounds the stem. Occurs in many monocots and certain dicots.

leaf trace. Vascular tissue that extends into the leaf from the stem. The portion of a vascular bundle extending from the stem to the base of a leaf. A single leaf may have one or more leaf traces. (*See* figs. M-10, P-15).

leaky gene. *See* hypomorphic gene.

lectotype. A specimen selected from the original collection of plant material, to serve as the nomenclatural type when the holotype is missing, or if it was not designated when the taxon was first published.

Fig. L-3. Leaf axil: A. leaf blade, B. petiole, C. leaf axil, D. axillary bud.

legume. **1.** A type of dry fruit developed from a single carpel and when mature, dehiscing along two sides (sutures). **2.** Any member of the pea family (Fabaceae), e.g. beans, peanuts and alfalfa.

lemma, *pl.* **lemmata.** In grasses, one of the pair of bracts (lemma and palea) that subtends the floret. The lemma is lower, usually larger, and frequently bears an awn, which is an extension of its midrib. (*See* fig. F-3).

lentic. A freshwater habitat characterized by standing (nonflowing) water, as in lakes, ponds, swamps or bogs. Contrast lotic.

lenticels. Spongy areas in the cork (phellum) surfaces of stems, roots, certain fruits (apples, plums, pears) and other plant parts which allow interchange of gases between internal tissues and the atmosphere. The first lenticels frequently arise under stomata. Their size varies from a few millimeters to several centimeters in length and may occur singly or in rows. (*See* fig. -B-8).

lenticular. Lens-shaped; double convex with two edges.

lepides. Scales.

lepidote. Covered with small, scurfy scales.

lepta-, lepto-. A prefix meaning small, thin, slender.

leptom, leptome. All of the conducting cells (sieve elements, companion cells) and parenchyma cells of the phloem, excluding supporting cells (fibers and sclereids). Contrast hadrom.

leptoma. A thin region of exine on a pollen grain from which the pollen tube usually emerges.

leptomorph. One of two general types of rhizomes, characterized by a slender stem, long internodes, and indeterminate growth; they spread extensively over an area rather than producing a clump; e.g. lily-of-the-valley (*Convallaria* spp.) Contrast pachymorph.

leptonema. *See* leptotene.

leptosporangiate ferns. Ferns having sporangia (leptosporangia) which arise from single cells of the epidermis; found in all but a few genera of ferns.

leptosporangium, *pl.* **leptosporangia.** A sporangium originating from a single superficial (epidermal) parent cell which first produces a stalk; the apical cell develops into the annulus and sporogenous cells. Relatively few spores are produced. Compare eusporangium. (*See* fig. E-5).

leptotene.   The first stage of prophase I of meiosis. The chromosomes become visible at this stage as long, thin, threadlike structures. *Syn.* leptonema and leptophase.

lethal mutant.   A mutation which will result in the premature death of the individual carrying it. Dominant lethals kill heterozygotes, recessive lethals kill only homozygotes.

leuco-.   A prefix meaning white.

leucoplast.   A colorless plastid contained in plant cells, functioning as storage bodies for such products as starch, oil and protein.

leucosin.   A polysaccharide food storage product composed of glucose units, occurring as granules and oil droplets in cells of yellow-green algae (Xanthophyta), diatoms and golden-brown algae (Chrysophyta).

liana.   A large, woody, tropical climbing plant or vine.

libriform fiber.   A xylem fiber commonly with thick walls and simple slitlike pits. Usually the longest cell in the wood.

lichen.   A composite structure consisting of a fungus (usually an ascomycete) and algal

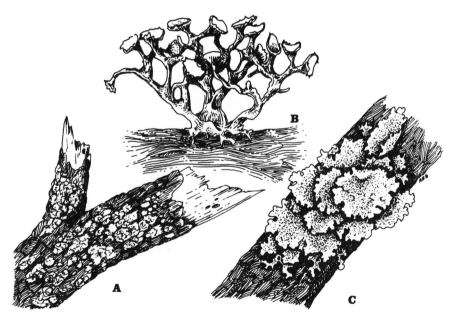

Fig. L-4. Lichens: A. crustose lichen, B. fruticose lichen, C. foliose lichen.

cells which are so closely associated as to form distinctive plantlike individuals. *See* crustose, foliose, fruticose for characteristic forms. (*See* fig. L-4).

lid. *See* operculum.

life cycle. In any organism, the entire sequence of events in its growth and development from the time of zygote formation until the time of gamete formation.

life form. The habit or growth form of a plant in relation to its environment, usually a constant characteristic but may vary under different environmental conditions.

light reactions. The reactions of photosynthesis in which light energy is required for activation.

ligneous. Woody.

lignicolous. Growing in wood; as with fungal mycelia.

lignification. The impregnation of cellulose with lignin.

lignin. A complex organic substance derived from phenylpropane and distinct from carbohydrates. It is an important ingredient of the material surrounding the cellulose microfibrils in the secondary wall of many plant cells. It imparts rigidity and strength especially to woody tissues.

ligulate. With a ligule; strap or tongue-shaped, as a petal or leaf.

ligule. **1.** A strap-shaped structure. **2.** A small membraneous structure or ring of hairs projecting from the top of the leaf sheath in many grasses (Poaceae), (*See* fig. A-22) and some sedges (Cyperaceae). **3.** A strap-shaped corolla, as in the ray flowers of composites (Asteraceae). **4.** A tongue-shaped outgrowth on the adaxial (upper) surface of the leaves (microphylls) and sporophylls of *Isoetes* and *Selaginella*. (*See* fig. M-6).

limb. **1.** A large tree branch. **2.** The upper, expanded portion of a united corolla or calyx above the tube, throat or claw.

lime, liming. Calcium carbonate ($CaCO_3$). Also the practice of adding $CaCO_3$ to the soil to help correct an acidic condition.

limiting factor. Any physical or chemical factor whose deficiency or excess is deleterious to living organisms.

limnology. The study of fresh water biology.

linear. Long and narrow, the sides parallel or nearly so; usually more than ten times longer than broad.

linear tetrad. An arrangement of four megaspores in a single row. A linear triad has three megaspores in a row. *See* megaspore.

lineate. Marked with parallel lines.

lineolate. Marked with fine lines.

lingulate. Tongue-shaped.

linkage, linkage groups. The tendency for two or more genes to be inherited together because they are located on the same chromosome.

linkage map. Delineation of the relative position of genes on chromosomes as determined by genetic recombination studies.

linters. The short fibers remaining on cotton seed after ginning.

lip. **1.** One of the two divisions of a bilabiate corolla or calyx, i.e. one in an upper (superior), and one in a lower (inferior) portion. **2.** The labellum of orchids.

lip cells. The two, thin-walled, slightly enlarged cells in the front of many fern sporangia; through action of the annulus, the sporangium ruptures along a line between the lip cells to discharge the spores. (*See* fig. L-5).

lipid. A general term for compounds containing fat, such as true fats, oils, waxes, phospholipids, sterols, and carotenes; these are insoluble in water but soluble in organic solvents (e.g., alcohol, benzene, or ether).

liter. The metric standard measure of volume, equal to one cubic decimeter.

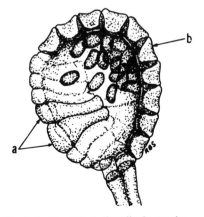

Fig. L-5. Lip cells: a. lip cells, b. annulus.

lithocyst. A cell containing a cystolith.

lithotroph. An organism that manufactures food by utilizing energy derived from the oxidation of inorganic materials such as hydrogen sulfide or ammonia.

litter. The layer of undecomposed or partly decomposed plant debris on the surface of the ground. *Syn.* duff.

littoral (zone). **1.** Of or pertaining to the seashore, especially the region between high and low tides. **2.** In lakes, pertaining to the shallow water region in which light is able to penetrate to the bottom; typically occupied by rooted plants.

liverwort. *See* bryophyte.

livid. Pale lead-colored.

loam. A soil texture composed of approximately 10 percent clay, 45 percent sand, and 45 percent silt. When it contains a substantial proportion of organic matter, it becomes an excellent agricultural soil because it absorbs water readily, permits easy root penetration, and has good drainage and aeration.

lobate. Divided into lobes.

lobe, lobed. Any division or segment of an organ (usually rounded); specifically, a part of a leaf, petal, or calyx cut less than halfway to the center.

lobulate. Divided into small lobes.

lobule. A small lobe.

-locular, -loculate. A suffix meaning to have compartments, chambers or cavities; divided into locules. A bilocular structure has two locules, a trilocular structure has three, etc.

locule. **1.** A compartment, chamber or cavity; as in an ovary, anther or fruit. (*See* fig. M-6). **2.** In Ascomycetes, the stromatic chamber containing asci. *Syn.* loculus.

loculicidal. Dehiscence or splitting along the walls of the locules of a capsule, as contrasted to dehiscence along the septae. Compare septicidal.

locus, *pl.* **loci.** The position of a gene on a chromosome.

lodging. A condition found among cereal plants in which heavy fertilization with nitrogen-containing fertilizers causes the plants to greatly elongate and thus fall over easily.

lodicule.   One of two or three minute hyaline scales at the base of the stamens of most grasses (Poaceae), believed to be rudiments of ancestral perianth parts. (*See* fig. F-3).

loment.   A modified legume fruit having constrictions between the seeds, which at maturity, separates at the constrictions into one-seeded segments.

long-day plant.   Plants which flower only after receiving illumination longer than a "critical photoperiod", which varies in length (approximately 11-14 hours), depending on the species. Compare short-day plant.

longitudinal section.   A section cut lengthwise, along the long axis.

long shoots.   In certain woody plants, (e.g. *Ginkgo* and *Pinus*), branches that increase rapidly in length during the first season emerging from the bud. They have widely separated nodes that bear specialized lateral spur shoots (or short shoots) from which arise a cluster of leaves or reproductive organs. (*See* fig. L-6).

lorate.   Strap-shaped, ligulate; often also flexuous with the margins wavy, as in certain leaves.

Fig. L-6. Long shoot: A. immature cone, B. leaf fascicle, C. mature open cone.

lorica. A surrounding case or shell which is separate from and encloses the protoplast (cell) of some algae, e.g. certain euglenoids (Euglenophyta) and golden-brown algae (Chrysophyta).

lotic. A freshwater habitat characterized by running (flowing) water, as in rivers, streams and springs. Contrast lentic.

lumen. The space within a plant cell bounded by the cell wall; the cavity left within a cell after the protoplast dies, as in tracheids.

lunate. Crescent-shaped; half-moon shaped. *Syn.* selenoid.

lurid. Dingy, yellowed.

lutein. A yellowish pigment, the most abundant xanthophyll found in the leaves of plants.

lux. A metric unit of illumination equal to 0.0929 footcandles; one footcandle equals 10.76 lux.

lycopod. A group of seedless vascular plants of which the main genera are *Lycopodium* (club mosses), *Selaginella* (spike mosses) and *Isoetes* (quillworts).

lyrate. Pinnatifid, but with a large, rounded terminal lobe, and smaller lower lobes; lyre-shaped.

lysigenous. Referring to intercellular spaces which are formed by the dissolution of cells, as lysigenous canals. Compare rexigenous and schizogenous.

lysimeter. Instruments used to measure the amount of water gained and lost from the soil. Essentially, they are large or small soil containers which are hydrologically isolated from the surrounding soil. *Syn.* evapotranspiration gauge.

-lysis, -lytic, -lyte. A suffix meaning to break apart, to loosen, or to dissolve.

lysis. Disintegration, as of a compound by an enzyme; the breakdown of a bacterial cell membrane by a phage.

lysogenic bacteria. A bacterial cell which carries a phage (virus) from one generation to another in a noninfectious form. Eventually the phage enters an active cycle of infection, resulting in lysis of their bacterial hosts.

lysogeny. The integration of the genetic material of a virus and its bacterial host.

lysosome. A cellular organelle bounded by a single membrane, which contains acid hydrolytic enzymes capable of breaking down proteins and other complex organic molecules; the enzymes are also capable of destroying the cell in which the lysosome is located.

M

mace. A spice derived from the aril of nutmeg.

macro-. A prefix meaning large or long.

macroconidium, *pl.* **macroconidia.** A multinucleate conidia of normal size produced by certain ascomycetous molds (e.g. *Neurospora*) which also produce uninucleate microconidia.

macroconsumer. *See* phagotroph.

macrocyclic. Long-cycled; applied to species of rust that produce one or more types of binucleate spores in addition to teliospores.

macrocyst. In Myxomycetes, the inactive form of the plasmodium in which multinucleate portions of the sclerotium have been separated by wall material. Also, walled cell masses produced by the cellular slime molds (Acrasiomycetes).

macrofibril. An aggregation of microfibrils, visible with the light microscope. (*See* fig. M-5).

macrogamete. *See* egg.

macromolecule. A very large molecule with high molecular weight; e. g. proteins, polysaccharides and nucleic acids.

macromutations. Gene mutations that produce a phenotype which more closely resembles distantly related species or genera than it does the parental species.

macronutrient. A mineral required in relatively large amounts for plant growth, as nitrogen, phosphorus, potassium, calcium, magnesium or sulfur.

macrophyll. *See* megaphyll.

macrosclereid. An elongated sclereid with unevenly distributed secondary wall thickenings; common in seed epidermis of legumes (Fabaceae). *Syn.* malpighian cell. (*See* fig. S-3).

macroscopic. Visible to the unaided eye.

macrosporangium. The structure in which macrospores are produced. *See* megasporangium.

macrospore. The larger of the two kinds of spores, as in *Selaginella* and related plants. *See* megaspore.

maculate. Spotted or blotched.

malacophily. Pollination by snails or slugs.

male sterility. A condition in plants in which no viable pollen is produced.

male symbol ♂. The shield and spear, the zodiac sign for Mars.

malpighiaceous hairs. Straight appressed hairs, attached by the middle and tapering to the free tips.

malpighian cell. *See* macrosclereid.

malt. Germinated barley or other grains (e.g. rice and maize) which are dried and ground. Used in the manufacture of many beers and other products.

maltose. Malt sugar.

mammiform, mammilliform, mammose. Breast-shaped, conical with rounded apex.

mammillate. Having nipples or teat-shaped processes.

manicate. Having such a thick and interwoven pubescence that it can be stripped off like a sleeve.

mannitol. An alcohol storage product of brown algae (Phaeophyta).

manubrium. 1. A prismatic cell bearing capitula and antheridial filaments in some green algae (e.g. Charophyta). 2. A long, thin, cylindrical base of some cymbas and spathes.

map unit. A unit that equals a recombination frequency of one percent.

marcescent. 1. Withering but not falling off, as corollas and stamens in certain kinds of flowers. 2. Leaves of short duration, dying at the end of the growing season.

marginal growth. The growth along the margins of a leaf primordium that results in the formation of the leaf blade.

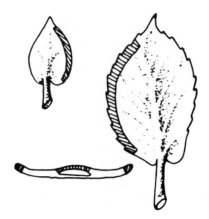

Fig. M-1. Marginal meristem.

marginal initials.   Cells along the margins of a growing leaf that contribute cells to the protoderm.

marginal meristem.   The meristem located along the margin of a leaf primordium, contributing to the marginal growth of the leaf blade. (*See* fig. M-1).

marginate.   Having a distinct margin.

margo.   The membrane which surrounds the torus in a bordered pit-pair of gymnosperm tracheids; consists of bundles of microfibrils. *Syn.* pit membrane. (*See* fig. B-6).

marine.   Growing in the ocean, or influenced by it.

maritime.   Pertaining to the sea.

marsupium.   The subterranean, usually rhizomatous fruiting receptacle of certain liverworts.

mass meristem.   Meristematic tissue in which the cells divide in various planes so that the tissue increases in volume. *Syn.* block meristem.

mass number.   The number of protons and neutrons in the atomic nucleus. Symbol, A.

massula, *pl.* **massulae.**   1. A segment of the periplasmodium of *Azolla* made up of a mass of microspores in a sporangium, embedded in a mucilaginous matrix. 2. A clump of microspores, as in certain orchids.

mastigonemes.   The hairlike or threadlike processes which occur on the flagella of a number of eukaryotic cells; flagella bearing mastigonemes are commonly referred to as tinsel flagella.

maternal inheritance. Phenotypic differences between identical genotypes due to the influence of maternal cytoplasm.

mating system. The method of pollination in a specific plant; e. g. self-pollination, cross-pollination, or a combination of both.

mating type. Morphologically similar individuals that differ in their mating behavior. Such individuals have surface proteins that will only bind to complementary proteins or polysaccharides found on the coats of individuals of opposite mating types. Found in some algae and fungi.

matrix. A complex, structured, semisolid substance in which something is embedded.

matrix potential. The ability of water to move through or into a substance by capillary action; as water moving through dry soil, or into a seed.

maturation region. *See* region of maturation. (*See* fig. M-2).

matutinal. Having flowers that open early in the morning.

mazaedium, *pl.* **mazaedia.** A fungal structure which produces spores that form a powdery mass when liberated from the ascus.

mealy. *See* farinaceous.

mean. The arithmetic average of a series of observations. *Syn.* average.

mechanical tissue. *See* supporting tissue.

median. Value of a series of observations, on each side of which there is an equal number of larger and smaller variates.

medium, *pl.* **media.** **1.** A liquid or solid organic substrate (e.g. agar) upon which many types of organisms (especially microorganisms) can be grown or cultured. **2.** In horticulture and agriculture, various materials and mixtures which are used as a medium in which to propagate plants by seed or cuttings; e. g. sand, soil, vermiculite, etc.

medulla. **1.** The central layer of a lichen thallus composed of loosely arranged fungal hyphae. **2.** The central portion of the stipe and lamina of brown algae (Phaeophyta). **3.** The pith of a stem.

medullary. Of, or pertaining to the pith of a stem, or to the medulla.

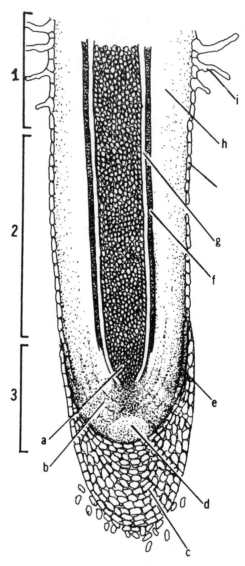

Fig. M-2. Maturation zone: 1. region of maturation, 2. region of elongation, 3. region of cell division; a. procambium, b. ground meristem, c. root cap, d. apical meristem, e. protoderm, f. endodermis, g. pericycle, h. cortex, i. root hair.

medullary bundles.   Vascular bundles located in the pith region of a stem, frequently associated with anomalous secondary growth; occurs in many dicotyledonous families.

medullary ray. *See* interfascicular region.

medullary sheath. *See* perimedullary region.

mega-. A prefix meaning large.

megafossil (plant). Large fossils such as wood, leaves, fruits and seeds. Compare microfossil.

megagamete. *See* egg.

megagametophyte. The female gametophyte which develops from the megaspore and produces the female gamete(s).

megaphyll. The type of leaves found in ferns, gymnosperms, and angiosperms which have complex venation patterns, and usually have a leaf trace associated with a leaf gap. *Syn.* macrophyll. *See* Telome theory and microphyll.

megasporangium. The sporangium in which megaspores are produced; the nucellus of seed plants. *Syn.* macrosporangium.

megaspore. In heterosporus plants, the haploid spore(s) produced by meiosis of the megasporocyte, that gives rise to the female gametophyte (megagametophyte). In seed plants, the megaspore develops within the megasporangium. *Syn.* gynospore.

megaspore mother cell. *See* megasporocyte.

megasporocyte. A diploid cell in the megasporangium which undergoes meiosis producing four haploid megaspores, of which only one usually survives to become the functional megaspore. *Syn.* megaspore mother cell; meiocyte.

megasporogenesis. The developmental stages through which megaspores are produced.

megasporophyll. **1.** A leaf (e. g. *Isoetes*) or leaflike structure (e. g. *Cycas* or *Selaginella*) that bears one or more megasporangia or ovules. **2.** In angiosperms, a carpel.

megastrobilus. The female cone of many types of gymnosperms consisting of megasporophylls; found in cycads, conifers and *Welwitchia*. (*See* fig. M-3).

meiocyte. *See* megasporocyte.

meiosis. A fundamental process of sexual reproduction by which a diploid (2n) organism produces haploid (1n) gametes or spores. During interphase, chromosome (DNA) duplication occurs. Meiosis then follows with two successive nuclear divisions resulting in the production of four haploid cells from one diploid cell. *Syn.* reduction division.

Fig. M-3. Megastrobilus cone of a cycad.

meiosporangium.   A sporangium in which spores are produced by meiosis, as in certain fungi, e.g. chytrids.

meiospore.   A haploid spore produced by meiosis. (*See* fig. A-9).

meiotic drive.   Any meiotic mechanism which results in the unequal recovery (via fertilization) of only certain functional megaspores produced by a heterozygote.

melittophily.   Pollination by bees.

membranaceous, membranous.   Thin, soft, flexible, and more or less translucent.

Mendel's principle of dominance.   States that one gene of a pair (alleles) may mask or inhibit the expression of the other member of the pair. In actuality, dominance does not occur between all gene pairs.

Mendel's principle of independent assortment.   States that different pairs of genes are inherited independently of one another assuming that they occur on separate chromosomes. This inheritance is governed by chance. Also known as Mendel's Second Law.

Mendel's principle of segregation.   States that the two genes of a pair separate during meiosis so that each gamete has only one member of each pair. Also known as Mendel's First Law.

Mendel's principle of unit characters.   States that organismic traits are inherited and this inheritance is controlled by factors called genes which occur in pairs.

meniscoidal.   Thin and concave-convex.

meniscus.   The curved upper surface of a liquid column.

mentum.   An extension of the base of the column in the flower of some orchids.

mericarp.   One of the two seedlike carpels of the fruit of the umbel family (Apicaceae).

meridional.   Occurring at right angles to an equatorial plate.

meristele.   An individual vascular bundle of the dictyostele of certain plants (e.g. ferns and *Selaginella*), variable in shape and consisting of primary phloem enclosed by primary xylem.

meristem.   A tissue in which mitotic cell divisions occur, generally localized within the plant body and constituting a permanent or temporary zone of actively dividing cells from which mature tissues differentiate. *See* apical meristem.

meristematic.   Of, pertaining to, or arising from a meristem.

meristematic variation.   Variation in numbers of parts or organs.

meromixis.   In bacteria, genetic exchange involving a unidirectional transfer.

merosity.   The absolute number of parts within a floral whorl, as the number of petals in the corolla.

merosporangium, *pl.* **merosporangia.**   Cylindrical sporangia which typically contain a single row of sporangiospores.

-merous.   A suffix indicating the number of parts; as flowers five-merous in which each of the floral organs present are in fives, e. g. petals, stamens or sepals.

Mertesian mimicry.   When one deadly and one harmless organism both mimic a moderately poisonous or distasteful model.

mesarch xylem.   A pattern of maturation of primary xylem from the center outward, so that the oldest xylem (protoxylem) is surrounded by younger metaxylem; common in ferns and infrequent elsewhere.

mesic.   Moist, especially in reference to habitats or environments. Compare xeric.

meso-.   A prefix meaning middle or intermediate.

mesocarp.   The middle layer of cells of the pericarp or fruit wall. Between the exocarp and endocarp.

mesocotyl.   The internode separating the coleoptile and the scutellar node in the embryo and seedling of grasses (Poaceae).

mesogamy.   A method of ovule penetration in seed plants in which the pollen tube laterally penetrates the integument or funiculus before reaching the egg apparatus of the embryo; e.g. occurs in *Alchemilla, Cucurbita* and *Circaeaster*. Compare porogamy.

mesokaryote.   Organisms, particularly dinoflagellates (Pyrrophyta), which because of chromosomal characteristics appear intermediate between prokaryotes and eukaryotes. Their chromosomes contain no RNA, histones or other proteins, remain condensed and recognizable throughout the cell cycle, and have a fibrillar organization (in some species).

mesome.   According to the Telome theory, a portion (''internode'') of the stem or of the rootlike axis between successive branches of a primordial vascular plant.

mesomorphic.   Referring to the structural features of mesophytes.

mesophyll.   The photosynthetic tissue between the upper and lower epidermis of a leaf; mesophyll may be homogenous or differentiated into palisade and spongy parenchyma layers. (*See* figs. H-6, M-4).

mesophyte, mesophytic.   Plants which grow in environments that are neither very wet nor very dry; plants intermediate in moisture requirements between hydrophytes and xerophytes.

mesosome.   Membraneous bodies attached to the plasmalemma in prokaryotic cells.

mesozoic era.   A geologic era extending from 65 to 230 million years ago, during which many of the major categories of organisms now present on earth evolved.

messenger RNA, mRNA.   A form of RNA produced in the nucleus and reflecting the base sequence of DNA. After transcription it moves into the cytoplasm and comes into contact with the ribosomes where it determines the sequence of amino acids in a polypeptide.

mestome sheath.   The thick-walled, inner bundle sheath enclosing the vascular tissues in the leaves of some grasses (e.g. Poaceae, subfamily Festucoideae); may have suberized lamella in its walls and be analogous to an endodermis.

metabolic pathway.   A series of consecutive enzymatic reactions resulting in the conversion of one molecule to another.

metabolism.   The sum total of all chemical processes occurring within a living cell.

metabolite.   A substance involved in metabolism.

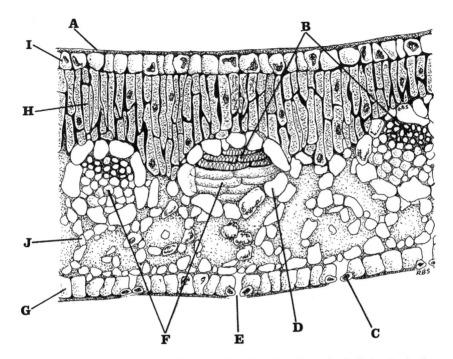

Fig. M-4. Mesophyll in leaf cross section: A. cuticle, B. xylem, C. guard cell, D. bundle sheath, E. stoma, F. phloem, G. lower epidermis, H. palisade parenchyma, I. upper epidermis, J. spongy parenchyma.

metaboly.   The ability of an organism to change or alter its external body form, as with many *Euglena* species.

metacentric.   A chromosome which has a centrally located centromere.

metalloenzyme.   A metal-containing enzyme.

metalloporphyrin.   Metal-containing compounds including chlorophyll, cytochrome, and heme.

metaphase.   A phase of mitosis or meiosis during which the chromosomes assemble more or less midway between the two poles of the spindle.

metaphloem.   Primary phloem that differentiates after the protophloem has matured, and before secondary phloem is formed. In plants without secondary growth, metaphloem is the only conducting phloem in mature plant parts.

metaphyll.   An adult leaf.

metaxenia.   The influence that a male plant has, through its pollen, on fruit development of the female plant; effects include amount of fruit set, size of fruit, and date of ripening; occurs in dates, apples, oaks, cotton, etc. Compare xenia.

metaxylem.   Primary xylem that differentiates after the protoxylem has matured, and before the secondary xylem is formed (if any, depending on the taxon). Matures largely after the stem or root section has completed its elongation.

micelle.   **1.** Portions of cellulosic microfibrils in which the cellulose molecules are arranged in a crystalline lattice structure. (*See* fig. M-5). **2.** In colloidal systems, an individual particle dispersed in a liquid medium.

micro-.   A prefix meaning small.

microbe.   A microorganism.

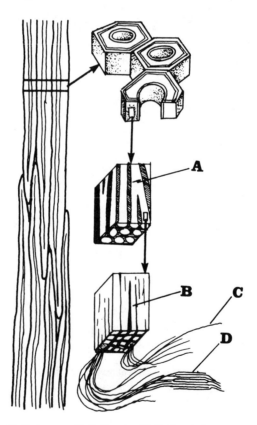

Fig. M-5.   Microfibril: A. macrofibril, B. microfibril, C. cellulose molecule, D. micelle.

microbiology. The study of microorganisms.

microbody, microbodies. Subcellular organelles containing specialized enzymes with specific metabolic functions, such as photorespiration and fat metabolism. Structurally, they appear bounded by a single unit membrane. *See* peroxisome and glyoxysome.

microclimate. The climate in the immediate vicinity of an organism.

microconidium, *pl.* **microconidia.** The smaller of two conidia (the larger is a macroconidia) produced by some fungi (e.g. *Neurospora*) that often acts as a spermatium.

microcyclic. Short-cycled; applied to species of rusts in which the teliospore is the only binucleate spore.

microcyst. A small, encysted protoplast; usually an encysted myxamoeba of the plasmodial slime molds (Myxomycetes) or cellular slime molds (Acrasiomycetes).

microfibril. The primary structural components of a plant cell wall, composed of chain-like cellulose molecules; visible only with the electron microscope. *See* macrofibril. (*See* fig. M-5).

microfossil (plant). Microscopic fossils such as pollen grains, spores, algae, fungi or fossil fragments such as tracheids, pieces of cuticle, etc. Compare megafossil.

microgametophyte. The male gametophyte of a heterosporus plant which develops from the microspore and produces the male gametes; the pollen grain in a seed plant.

micromelittophily. Pollination by small bees.

micrometer. A unit of microscopic measurement abbreviated μm; 1/1000 of a millimeter, or 1/25,000 inch. *Syn.* micron.

micron. *See* micrometer.

micronutrient. A mineral required in relatively small amounts for plant growth, as iron, chlorine, copper, manganese, zinc, etc. *Syn.* trace element, micromineral elements.

microorganism. Organisms too small to be seen without a microscope.

microphyll. Relatively small leaves (except in *Isoetes* and certain extinct lycopods) with simple venation consisting of a single vascular strand; they do not form a leaf gap at the node of the stem. Also found in extant lycopods. *See* megaphyll and enation theory.

microphyllous. Having microphylls.

micropyle.   A small opening in the integuments of an ovule through which the pollen tube usually enters to reach the nucellus (megasporangium). (*See* fig. O-3).

microspecies.   Sometimes applied by botanists to delimit a more or less uniform population which is slightly different morphologically from related uniform populations.

microsporangium.   In heterosporous plants, the sporangium in which microspores are produced; in angiosperms, the anther locule and its walls. *Syn.* androsporangium. (*See* fig. M-6).

microspore.   A haploid spore produced by meiosis of the microsporocyte and developing into the male gametophyte; usually, but not necessarily smaller than the megaspore of the species; the pollen grain of seed plants.

microspore mother cell.   *See* microsporocyte.

microsporocyte.   A diploid cell in the microsporangium which undergoes meiosis, producing four microspores; termed pollen mother cell in seed plants. *Syn.* microspore mother cell. (*See* fig. T-1).

microsporogenesis.   The developmental stages through which microspores are produced.

microsporophyll.   **1.** A leaf (e.g. *Isoetes*), or leaflike structure that bears one or more microsporangia. (*See* fig. M-6). **2.** In angiosperms, the stamen.

microstrobilus.   The male cone of many types of gymnosperms consisting of microsporophylls; found in cycads, ginkgo, conifers, *Welwitchia*, etc.

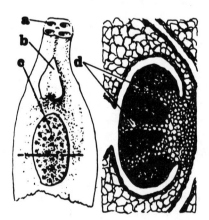

Fig. M-6. Microsporophyll: a. air chamber, b. ligule, c. microsporangium, d. trabeculae.

microtome.   An instrument designed for cutting very thin tissue sections for study with light or electron microscopes.

microtubules.   Slender proteinaceous tubules of indefinite length associated with many types of eukaryotic cells; the basic unit of structure of all eukaryotic cilia and flagella. Functions include formation of the mitotic and meiotic spindle fibers and phragmoplasts of dividing cells, participation in cell wall formation, protoplasmic streaming and providing a pathway for the entrance of solutes into the cell.

middle lamella.   An intercellular substance, composed mostly of pectic compounds, that cements the primary walls of contiguous cells. *Syn.* intercellular substance, intercellular layer.

midrib.   The main or central rib or vein of a leaf.

miliaris.   A minute, glandular spot on the epidermis.

milli-.   A prefix meaning one-thousandth.

millicurie.   One-thousandth of a curie; the amount of radioactive material in which 3.7 $\times$ 10^7 disintegrations occur per second.

milliliter (ml).   One-thousandth of a liter; the volume found in a cube one centimeter on a side. One milliliter of water weighs one gram at $4°C$.

millimeter (mm).   One-thousandth of a meter.

millimicron.   *See* nanometer.

mimicry, *pl.* **mimicries.**   Any situation where resemblance of one organism, to another organism or to an inanimate object(s), results in increased potential for survival (i.e., results in increased Darwinian fitness). In plants mimicry may involve only part of the organism. Examples in the plant kingdom include floral, seed, vegetative and chemical mimicries. *See* Batesian, Mullerian and Mertesian mimicry. (*See* fig. M-7).

mispairing.   A DNA double helix in which a nucleotide in one chain does not have a complementary nucleotide in the other chain.

mitochondrion, *pl.* **mitochondria.**   A double-membrane bound, cytoplasmic organelle of eukaryotic plants, in which cellular respiration occurs; the major source of ATP in nonphotosynthetic cells.

mitogametes.   Gametes produced by mitosis in haploid (1n) plants.

mitosis.   Nuclear division in which chromosomes duplicate and divide to yield two

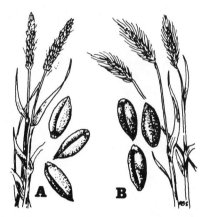

Fig. M-7. Mimicry involving seeds: A. rye plant and seeds, B. wheat plant and seeds.

"daughter" nuclei that are genetically identical to the original nucleus. Divided into four phases; prophase, metaphase, anaphase and telophase. Mitosis is usually followed by cell division (cytokinesis). *Syn.* karyokinesis.

mitosporangium, *pl.* mitosporangia.   A sporangium in which spores (mitospores) are produced by mitosis, rather than meiosis; as found in certain chytrids (*Allomyces*).

mitospore.   A haploid spore produced by mitosis in a mitosporangium.

mitotic index.   The fraction of cells undergoing mitosis in a particular sample.

mitotic recombination.   Somatic crossing-over.

mixed buds.   Buds that produce both leaves and flowers.

mixed sorus.   A fern sorus in which the sporangia develop at different rates, and thus mature at different times.

mode.   The value in a series of observations which occurs in the greatest frequency.

modifier gene.   A gene which changes the phenotypic expression of a nonallelic gene.

molal solution.   A solution which contains one mole of a solute dissolved in one liter of water.

molar solution.   A solution containing Avogadro's number of molecules of dissolved solute in sufficient water to make one liter of the solution.

mole. A gram-molecular weight. The number of particles in one mole of any substance is always the same: $6,022 \times 10^{23}$.

molecular weight. The sum of the atomic weights of all atoms in a given molecule.

molecule. The smallest whole unit of a compound consisting of two or more atoms of the same or different elements.

monad. Pollen grains occurring singly, instead of united with other grains.

monadelphous. Stamens united by connation of their filaments into a single group, forming a tube or column; as in mallows (Malvaceae) and some legumes (Fabaceae). *See* synema.

monarch. Having a single protoxylem group.

moniliform. Necklace-shaped; like a string of beads; cylindrical, with contractions at regular intervals.

mono-. A prefix meaning one, once or single.

monocarpic. A perennial or annual blooming and fruiting only once and then dying.

monocentric (thallus). The growth form of certain chytrid fungi (Chytridiales) in which only a single sporangium is formed. Contrast polycentric.

monocephalic, monocephalous. Bearing only one head; scapose, as in a dandelion.

monochasium. A cyme reduced to single flowers on each axis.

monochlamydeous. Having a perianth composed of a single envelope or whorl.

monochromatic light. Light of a single wave length.

monoclinous. Having both stamens and pistils in the same flower. *Syn.* hermaphroditic, perfect.

monocolpate. Pollen grains with a single furrow or groove; characteristic of most gymnosperms, monocotyledons, and certain families in the order Ranales. (*See* fig. M-8).

monocot. An abbreviated term for monocotyledon.

monocotyledon. A member of the Monocotyledoneae, one of the two taxa (classes or subclass) of angiosperms, characterized by the following: one seed leaf (cotyledon),

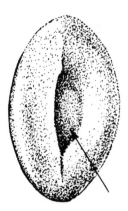

Fig. M-8. Monocolpate.

flower parts usually in three's (or multiples thereof), leaves often with parallel venation, root systems arising adventitiously and usually diffuse, and usually lacking secondary growth. Compare dicotyledon.

monoculture.   One-crop agricultural systems; these are inherently unstable ecosystems because of their simplicity.

monocyclic.   One-whorled, as in reference to floral parts.

monoecious.   A plant with separate male and female reproductive structures occurring on the same plant. In angiosperms, having unisexual flowers of different sexes on the same plant.

monogenic character.   A character determined by a single gene.

monograph.   A comprehensive treatise on a single taxonomic unit, such as a family or genus.

monogynous.   With one pistil.

monohybrid.   A cross between two parents differing by only a single genetic trait.

monokaryon, monokaryotic.   A condition in the hyphae of Basidiomycetes (club fungi) in which each hyphal segment contains a single nucleus.

monolectic.   *See* monotrophic.

monomer.   A subunit of a polymer; a simple molecule that can be linked with others like it to form a polymer.

monomorphic. **1.** Producing spores of only one form or type; e.g. zoospores. **2.** Having the same size and shape.

monopetalous. Gamopetalous.

monophyletic. The origin of organisms from a single common ancestor. Compare polyphyletic.

monophyllic. A flowering plant that is pollinated by one, or a very few related taxa of vectors.

monophyllous. One-leaved.

monoploid. *See* haploid.

monopodial. A form of branching, in contrast to dichotomy, in which lateral branches, if present, usually originate at some distance from the apex of the main axis. Occurs in some ferns, *Equisetum,* and most seed plants.

monopolar germination. Spore germination by the production of a germ tube at only one end.

monosaccharide. A sugar that cannot be broken down into smaller sugar molecules, e.g. glucose.

monosiphonous. Pollen grains which form only one pollen tube upon germination; typical of the majority of angiosperms. Compare polysiphonous.

monosome, monosomy. An organism which lacks one chromosome of its diploid complement, hence, having $2n-1$ chromosomes.

monosporangium. A sporangium that develops in a filament from differentiation of a vegetative cell thereby producing a single monospore; found in certain red algae (Rhodophyta), e.g. *Bangia* and *Porphyra.*

monospore. The single spore produced in a monosporangium of certain red algae (Rhodophyta); when released, they germinate and grow into new plants.

monosporic embryo sac. One of three major types of embryo sac development in angiosperms, in which the embryo sac is derived from a single megaspore.

monostichous. Arranged in a single vertical row, or along one side.

monostromatic. Having a thallus one cell thick as in the "leaves" of mosses, and the thalli of certain algae; e.g. *Monostroma,* a green alga.

monotrophic.   Animal pollinators (primarily solitary bees) which visit only one species of flower and collect both pollen and/or nectar. *Syn.* monolectic.

monotypic.   In taxonomy, having a particular taxon represented by only one subordinate member; e. g. a family with only one genus, or a genus with only one species.

montane.   Pertaining to mountains.

morph-, -morph.   **1.** Prefix or suffix meaning form. **2.** As a noun, indicates an individual.

morphogenesis.   The development of form and structure; includes the physiological and morphological changes that accompany the external and internal development of an organism.

morphology (plant).   The study of form, structure and development of plants.

mosaic.   **1.** With a variegated or mottled appearance, often caused by a viral infection; as found in certain leaves. **2.** An organism having portions of its cells or tissues genetically different from the rest of the individual.

moss.   *See* bryophyte.

mother cell.   *See* precursory cell.

moth-flowers.   Flowers that are pollinated by moths.

motile.   Capable of self-propulsion, usually by flagella.

motor cell.   *See* bulliform cells.

mottled.   Used to describe a blotched or spotted appearance.

mRNA.   *See* messenger RNA.

mucilage.   A gelatinous substance which absorbs water and increases in bulk; found in many types of plants and algae in various structures, e.g. stems, leaves, fruits, seeds.

mucilage canal, mucilage duct.   Cavities or canals in various plant tissues which contain gums or mucilages; e.g. found in cycad stems, certain brown and red algae, *Euglena,* etc.

mucilage cell.   A type of secretory cell which secretes or contains gums or mucilages, and often contains raphide crystals.

mucilaginous.   Having the character of, or containing mucilage; being gelatinous, gummy or sticky.

mucro.   A short, sharp, abrupt, terminal point.

mucronate.   Terminated abruptly by a distinct mucro.

mucronulate.   Diminutive of mucronate.

Mullerian mimicry.   A type of mimicry involving two or more species of poisonous, unpalatable, or otherwise harmful organisms that resemble one another.

multi-.   A prefix meaning several or many.

multiaxial.   A type of growth in which the main axis is composed of many parallel or almost parallel filaments; as in certain red algae (Rhodophyta). Compare uniaxial.

multicarpellate.   A compound ovary with two or more carpels. (*See* fig. M-9).

multiciliate.   With many cilia.

multicipital.   A rootstock or caudex from which several stems arise.

multicostal.   Many-ribbed.

multifid.   Divided into many narrow lobes or segments.

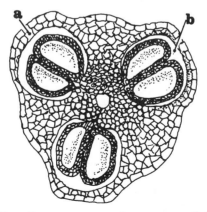

Fig. M-9. Multicarpellate ovary, cross section: a. ovary wall, b. locule.

Fig. M-10. Multilacunar node: a. median leaf trace, b. leaf gap.

multilacunar node.   A node which has numerous leaf gaps and leaf traces, associated with one leaf. (*See* fig. M-10).

multiperforate perforation plate.   A perforation plate which has many perforations.

multiple alleles.   Having many allelic forms of a single gene.

multiple epidermis.   A tissue composed of several cell layers, with only the outer layer having the typical epidermal characteristics; found in certain leaves and roots, possibly functioning in water storage. *See* velamen.

multiple fruit.   A fruit formed from the developed ovaries of several closely associated flowers which have a common axis; as in pineapple and osage orange. *Syn.* sorosis.

multiseriate.   Having many rows or many series.

multiseriate ray.   A ray in secondary vascular tissue which is several, to many cells wide.

multivalent.   An association of three or more partially paired chromosomes in meiosis.

muricate.   Rough, as a surface covered with many minute, sharp, protuberances.

muriculate.   Minutely muricate.

muriform.   Having bricklike markings, pits, or reticulations, as on some seed coats.

mushroom.   The common name of the fruiting body (basidiocarp) of certain basidiomycete fungi (Agaricales).

mutagen.   A substance or agent which causes mutations in chromosomes; e. g. certain chemicals, x-rays and various forms of radiation.

mutant.   A mutated gene or an organism carrying such a gene.

mutation.   An inheritable change in the genetic material of a cell; may be simple, involving one change in a single nucleotide pair, or complex, involving deletions, duplications, translocations, or inversions of chromosome segments.

muticous.   Pointless, blunt, awnless.

muton.   The smallest portion of a DNA molecule in which a mutation can occur; a single nucleotide.

mutualism, mutualistic.   A form of symbiosis in which both organisms derive benefit from the association.

myc-, mycet-, myco-.   A prefix meaning pertaining to fungi.

mycelium.   The collective term for the hyphae of a fungus; the vegetative thallus or body of a fungus.

mycobiont.   The fungal component of a lichen association.

mycologist.   One who studies fungi.

mycology.   The study of fungi.

mycoparasites.   Achlorophyllous angiosperms which obtain all water and nutrition from mycorrhizal associations with their roots.

mycophagy.   The eating of mushrooms.

mycoplasmas.   The smallest known prokaryotic organisms, ranging in size between 0.1 and 0.25 μ in diameter, and having no constant shape; mostly disease causing parasites of vascular plants, man, and cattle. *Syn.* pleuropneumonia-like organisms (PPLO).

mycorrhiza, *pl.* **mycorrhizae.**   A symbiotic, nonpathogenic association between fungi and roots of many types of plants; symbiosis may be mutualistic or somewhat parasitic; fungi may be superficial (ectotrophic), or contained within the host cells (endotrophic). *See* ectomycorrhiza, and endomycorrhiza. (*See* fig. M-11).

mycosis.   A fungus disease of man and other animals.

myophily.   Pollination by diptera; e.g. flies, gnats, mosquitoes or midges.

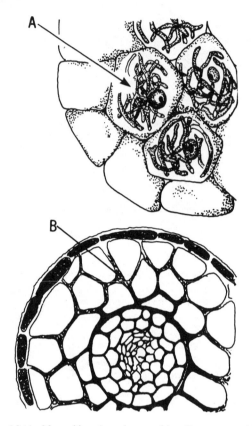

Fig. M-11. Mycorrhiza: A. endomycorrhiza, B. ectomycorrhiza.

myrmecophily.   Pollination by ants.

myrmecosymbiosis.   The mutualistic relationship between the ants and their host plant.

myxamoeba, *pl.* **myxamoebae.**   A haploid, amoeboid, nonflagellated cell, produced by meiosis in sporangia particularly of plasmodial and cellular slime molds (Myxomycetes and Acrasiomycetes); movement and appearance similar to true amoebae; may divide by fission and these cells may then function as gametes.

myxomycete.   A slime mold. Technically applicable to the true slime molds, Myxomycetes, but sometimes also applied to the cellular slime molds, Acrasiomycetes and Labyrinthulomycetes.

N

N. **1.** The haploid number of chromosomes; also written 1n. **2.** The number of individuals in a sample.

2N. The diploid number of chromosomes; also written 2n.

NAD. *See* nicotinamide adenine dinucleotide.

NADP. *See* nicotinamide adenine dinucleotide phosphate.

naked. Lacking its normal covering, or surrounding structures; nude; as in naked bud (a bud without bud scales) or naked flower (one without a perianth).

nannoplankton. Plankton with dimensions less than 70 to 75 microns.

nanometer (nm). One millionth of a millimeter; formerly termed millimicron.

napiform. Turnip-shaped.

nascent. In the process of being formed; as nascent or meristematic tissue.

nastic movements, nastic response. A growth response influenced by the internal organization of a plant and independent of the direction of external stimuli; e.g. photonasty, thermonasty and thigmonasty.

-nasty. Suffix indicating that a plant movement or response is determined by endogenous (internal) factors rather than external ones.

native. Indigenous. An individual or group of plants which was not introduced into a geographical area by man. Compare naturalized.

natural ecosystem. An ecosystem which develops in the absence of any major unnatural disturbance.

naturalized. An organism which is well established and reproducing in one area but originally came from another area. Compare native.

natural selection. A process of evolution which involves the interaction between an organism and its environment resulting in a differential rate of survival and reproduction of different genotypes in a population. The individuals best adapted to their environment are favored (selected) over the less adapted. Compare artificial selection.

natural system of classification. A system of classification based on phylogenetic (evolutionary) relationships.

navicular. Boat-shaped; shaped like the bow of a canoe.

neck. 1. The slender portion of the archegonium through which the male gamete enters. (*See* fig. A-18). **2.** That point where the blade and sheath separate in certain leaves. **3.** The constricted part of a corolla or calyx tube.

neck canal cells. Cells that fill the interior of an immature archeogonial neck, and disintegrate when the egg matures, leaving a moist canal through which the sperm passes. (*See* fig. A-18).

necrocoleopterophily. Pollination by carrion beetles.

necrosis. The death of cells, resulting either from injury or normal senescence.

nectar. A secretion from nectaries, containing up to 75 percent sugar, and sometimes also amino acids; acts as a pollinator attractant in angiosperm flowers, by attracting certain insects or birds; nectar is also an important source of food for these animals and others not involved in pollination.

nectar guides. Floral orientation cues directing a vector to the nectar. (*See* fig. N-1).

nectariferous. Nectar producing; having a nectary.

nectary. Any structure which secretes nectar, such as glands, trichomes or stomata-like orifices. Termed floral nectaries if associated with flowers or extrafloral nectaries if located on some other part of a plant.

needle. An elongated, slender leaf, as in pines (*Pinus*); pertaining to the shape of certain cellular crystals.

nema, *pl.* **nemata.** A filament; a thread.

nematodes (of plants). Elongate, nonsegmented worms of phylum Nematoda, which are important causative agents and vectors of many plant diseases; species affecting plants are small (0.5-2 mm long), migratory or sedentary, facultative or obligate parasites (mostly of roots), endo-, or ectoparasitic. *Syn.* roundworm, eelworm. (*See* fig. F-5).

neomorph. A mutant gene which produces a qualitatively new effect not produced by the normal allele.

neopolyploid. A recently derived polyploid, the parent(s) of which are extant.

neoteny. See paedomorphosis.

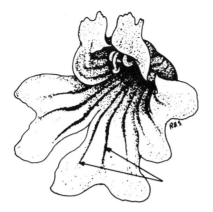

Fig. N-1. Nectar guides.

neotype. A specimen selected to serve as the nomenclatural type of a taxon when the original type specimen is known to have been destroyed.

neritic. A designation for that portion of the ocean adjoining the coast and extending out to a depth of about 200 meters. (*See* fig. A-16).

nervate, nervation. Having veins. *See* venation.

nerve. A simple vein or slender rib of a leaf or other organ.

nerved. Having a pronounced vein in a leaf, lemma, glume or other structure.

nervelet. An ultimate branch of a nerve.

net (apparent) photosynthesis. Net absorption of carbon dioxide in photosynthesis, without correction for carbon dioxide obtained from respiration within the tissue.

net primary production. The gross primary photosynthetic production minus the amount of chemical energy used for respiration.

net venation. The arrangement of veins in the leaf blade that resembles a net; characteristic of dicot leaves. *Syn.* reticulate venation.

neuter, neutral. Lacking sexual organs; sexless, as flowers lacking functional stamens and pistils.

neutron. An electrically neutral particle with a mass slightly greater than that of a proton found in the atomic nucleus of all elements except hydrogen.

nexine.   The inner, nonsculptured part of the exine of pollen grains. *Syn.* endexine. (*See* fig. S-6).

niche.   An n-dimensional hyperspace; defined as the place or position that a particular plant occupies in its ecosystem in regard to the utilization of its environment and its interactions with other organisms.

nicotinamide adenine dinucleotide (NAD).   A hydrogen acceptor molecule in respiration.

nicotinamide adenine dinucleotide phosphate (NADP).   A hydrogen acceptor molecule in photosynthesis.

nigrescent.   Blackish, or becoming black.

nitid.   Glossy; lustrous; smooth and clear.

nitrate ion (NO_3^-).   The primary source of nitrogen made available to plants under field conditions by the action of nitrifying organisms.

nitrification.   The conversion of ammonium or ammonia to nitrate, by certain bacteria or blue-greens. See nitrogen-fixing bacteria.

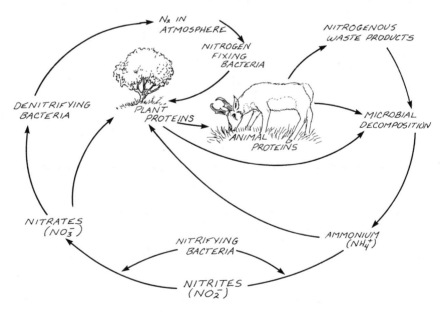

Fig. N-2. Nitrogen cycle.

nitrifying bacteria. *See* nitrogen-fixing bacteria. (*See* fig. N-2).

nitrogen base. A nitrogen-containing compound having the properties of a base (tendency to acquire a hydrogen atom); a purine (adenine or guanine) or pyrimidine (cytosine, thymine or uracil); a component of nucleotides.

nitrogen cycle. The worldwide circulation of the element nitrogen in various forms through the ecosphere. (*See* fig. N-2).

nitrogen fixation. The process by which gaseous nitrogen from the atmosphere is incorporated into organic nitrogen-containing compounds (nitrates) by the action of nitrogen-fixing bacteria, photosynthetic bacteria and certain free-living blue-greens.

nitrogen-fixing bacteria. Soil borne bacteria which are capable of converting ammonia and ammonium into nitrate through the process of nitrification; accomplished primarily by symbiotic forms as *Rhizobium* and also by free-living types, as *Azotobacter* and *Clostridium. Syn.* nitrifying bacteria. (*See* fig. N-2).

niveous. White; like snow.

nobel blade. A cultivation implement having a single wide blade that cuts just below the soil surface, destroying weeds without turning the soil. Used to preserve mulch on the surface without exposing moist soil to surface drying.

nocturnal. Occurring at night as in the opening of certain flowers.

nodal. Pertaining to the node.

nodal ring. A continuous ring of xylem and phloem at the nodes of the stem of *Equisetum* (horsetails).

nodding. Drooping, hanging down, pendent.

node. A region on the stem where leaves are attached; or the point of branching of the stem. Stems have nodes, roots do not. (*See* fig. P-15).

nodiform, nodose. Having knots or being knobby; usually in reference to roots, especially in the legumes (Fabaceae).

nodules. Enlargements or swellings on the roots of legumes (Fabaceae) and certain other plants inhabited by nitrogen-fixing bacteria.

nodulose. Having small knobs or knots.

nomenclature. The naming of things, particularly organisms, extant or extinct; in botany, refers to the correct usage of scientific names used in taxonomy.

nonarticulated laticifer.　A simple (single cell) laticifer originating as a single cell in the embryo; commonly multinucleate and may be branched or unbranched. Compare articulated laticifer. (*See* fig. L-2).

noncyclic reaction.　A chemical reaction in which an electron travels through a series of electron-acceptor molecules and does not return to its starting point.

nondisjunction.　The failure of homologous chromosomes to separate during meiosis or failure of paired chromosomes to separate during mitosis, resulting in the gametes either lacking or having extra chromosomes.

nonporous wood.　Wood in which the secondary xylem lacks vessel elements, as in conifers. (*See* fig. D-3).

nonsense mutation.　A mutant codon which does not code for an amino acid, thus no protein can be formed.

nonseptate.　Without cross-walls or septations.

nonstoried cambium.　Vascular cambium in which the fusiform initials and rays are overlapping, and are not arranged in horizontal rows as viewed on the tangential surface. (*See* fig. S-15).

nonstoried wood.　Wood formed from nonstoried cambia in which the axial cells and rays are not arranged in horizontal rows as viewed on the tangential surface.

nonvascular plants.　Plants which lack true vascular tissues of xylem and phloem, as bryophytes; traditionally may also include algae, slime molds, fungi and bacteria.

normal distribution.　A frequency distribution of variates normally found when they are influenced by a number of independent factors of about equal magnitude; results in a bell-shaped curve.

normalizing selection.　*See* stabilizing selection.

normal solution.　A solution which has a gram equivalent weight of solute dissolved in enough water to make a liter of solution.

notate.　Having spots or lines.

nototribic.　Flowers with stamens and/or stigmas turned so as to strike insect visitors on the dorsal (back) part of their body. Compare sternotribic.

novirame.　A flowering or fruiting shoot arising from a primocane, as in blackberries.

nucamentum.　An ament or catkin.

nucellar beak.   A conical protrusion of the nucellus that forms a cavity or chamber in which pollen may collect. Found in primitive seed plants.

nucellar cap.   A group of thick-walled and lignified cells which form a persistent hood over the apex of the embryo sac in certain angiosperms.

nucellus.   The tissue of an ovule, in which the female gametophyte (embryo sac) develops; the megasporangium. *Syn.* gynosporangium. *(See* fig. O-3).

nuciferous.   Bearing a nut.

nuclear body, nucleoid.   A structure in prokaryotes that contains the nuclear material and is embedded in a matrix distinct from cytoplasm.

nuclear envelope.   A perforated double membrane that surrounds the nucleus of a eukaryotic cell and is continuous with the endoplasmic reticulum.

nuclear pore.   A pore or perforation through the nuclear envelope, through which materials may pass between nucleus and cytoplasm. *(See* fig. N-3).

nuclear sap.   *See* nucleoplasm.

nucleated.   Having a nucleus.

nucleic acid.   DNA and RNA; large acidic compounds composed of linked nucleotides; they are concerned with the storage and replication of hereditary information and the synthesis of proteins.

nucleohistone.   A substance formed when DNA and histone are bonded together, as in chromosomes.

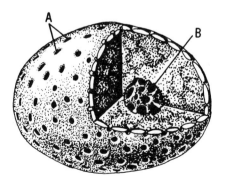

Fig. N-3. Nucleus: A. nuclear pores, B. nucleolus.

nucleolar organizer. A special region on some chromosomes associated with the formation of the nucleolus.

nucleolus, *pl.* **nucleoli.** One or more, roughly spherical shaped bodies present in the nucleus of eukaryotic cells; contain RNA and protein, and are the site of ribosome synthesis. (*See* fig. N-3).

nucleoplasm. In eukaryotic cells, the ground substance within the nucleus in which the chromosomes and nucleolus are suspended; in prokaryotic cells, the area that contains DNA. *Syn.* nuclear sap.

nucleoprotein. A compound composed of proteins and nucleic acids. There are two classes of proteins associated with DNA: protamines (low molecular weight) and histones (high molecular weight).

nucleoside. Nitrogenous base compounds in which pyrimidines and purines are linked to sugars (ribose or deoxyribose).

nucleotide. A structural unit of nucleic acids composed of an organic phosphate, a pentose sugar and a purine or pyrimidine (nitrogen base).

nucleus, *pl.* **nuclei.** An organelle present in all eukaryotic cells, enclosed in a nuclear envelope, containing the chromosomes, nucleoli and nucleoplasm. (*See* fig. N-3).

nucule. The female reproductive structure in the stoneworts (Charophyta); includes both the oogonium and the outer sterile protective cells.

nude. Naked or lacking its normal covering.

null hypothesis. An hypothesis which states that there is no discrepancy between observation and expectation based on some set of postulates.

nullisome. A diploid plant which lacks both members of a specific pair of chromosomes and has 2n-2 chromosomes.

numerical taxonomy. A field of taxonomy that employs a series of techniques that do not subjectively weigh any particular type of evidence for establishing relationships between taxa.

numerous. Usually meaning having more than ten of any structure.

nut. A one-seeded, dry, indehiscent fruit, with a hard or bony pericarp, derived from a simple or compound ovary, and characteristically, rich in oil; e. g. acorns, chestnuts or hazelnuts.

nutation.   A type of tropism in which certain plants exhibit more or less random motions, such as stem waving, in response to growth (cellular elongation), and/or other stimuli such as light. Such motions in vines cause them to grow around solid objects.

nutlet.   A small nut; one of a group of small seeds.

nutrient.   Any substance which promotes growth or provides energy for physiological processes.

nutrient cycling.   The pathway followed by nutrients from the physical environment to incorporation in living organisms and back to the physical environment.

nutriocyte.   The inflated portion of the ascogonium of some fungi (e.g. *Pericystis*), which eventually develops into an encysted spore.

nyctanthous, nyctigamous.   Flowering at night.

nyctinasty.   Endogenously (internally) controlled movements of plants which follow circadian rhythms, such as nocturnal folding of leaves, petal movement, formation of conidia in certain fungi, etc.

O

ob-.   A prefix usually meaning in the reversed direction or inverted position.

obcompressed.   Flattened at right angles to the primary plane or axis; compressed or flattened dorsiventrally.

obconic.   Inversely conical, with the point of attachment at the small end.

obcordate.   Inversely cordate with the notch at the apex. *Opp.* cordate.

obdiplostemonous.   Having stamens in two whorls, the outer opposite the petals and the inner opposite the sepals.

oblanceolate.   Inversely lanceolate; with the broadest portion nearest the apex and tapering toward the base.

oblate.   Nearly spherical but compressed at the poles.

obligate.   Essential, necessary; opposite of facultative.

obligate parasite.   A parasite that can develop only in living tissues, without a saprophytic stage.

oblique.   Slanted; with unequal sides; having an asymmetrical base.

oblong.   Much longer than broad with nearly parallel sides.

obovate.   Inversely ovate, with the terminal half broader than the basal.

obovoid.   Inversely ovoid, as a rounded object that is obovate in outline.

obscure.   Not clearly visible macroscopically; not distinct.

obsolescent.   Nearly obsolete or vestigial.

obsolete.   Vestigial, not evident, rudimentary; as an organ that is completely reduced or absent.

obturator.   A proliferation of placental tissue in certain angiosperm ovules which grows into the micropyle and forms a hood over the nucellus; facilitates entrance of pollen tube into ovule. (*See* fig. O-1).

242

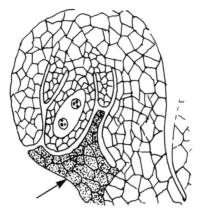

Fig. O-1. Obturator.

obtuse. Blunt or rounded at the apex, with the sides forming an angle of more than 90°; margins straight to convex.

occlusion. The healing of a wound in a tree by the growth of callus tissue.

oceanic. Of or pertaining to the open ocean. (*See* fig. A-16).

oceanic island. An island that has arisen from the sea floor usually as a result of volcanic action. Compare continental island.

ocellated. A broad spot of some color which has another spot of a different color contained within it.

ochroleucous. Yellowish-white.

ocrea, ochrea. A stipular sheath that encloses the base of a leaf, as in some members of the buckwheat family (Polygonaceae).

ocreate. Having a stipular sheath surrounding a stem above the insertion of a petiole or leaf.

ocreolate. Having smaller or secondary sheaths.

oculus. An eye or bud of a tuber (as a potato).

odd-pinnate. Pinnately compound with a terminal leaflet.

offset. A short, lateral shoot or branch which develops from the base of the main stem in certain plants, providing a means of asexual reproduction. *Syn.* offshoot, slip, sucker, ratoon.

offshoot. *See* offset.

-oid. A suffix meaning like or similar to.

oidiophore. A hyphal branch which produces oidia.

oidium, *pl.* **oidia.** A thin-walled, sporelike hyphal cell which acts like a spore or a spermatium. *See* arthrospores.

old-fields. Abandoned agricultural land.

oleosome. A membrane-bound organelle containing oil.

oligo-. A prefix meaning few.

oligolectic. *See* oligotrophy.

oligomerous. Having few parts in a given floral whorl.

oligophylic. A flowering plant which is pollinated by a few related taxa of vectors.

oligotaxy. Reduction in the number of whorls.

oligotrophic. Lakes or other bodies of water that are low in nutrients and organic matter.

oligotrophy, oligotrophic. Descriptive of insects that visit only a few species of plants for food. *Syn.* oligolectic.

omnivore. An organism that consumes both plant and animal food.

one-gene-one-polypeptide hypothesis. A hypothesis that a large class of genes exist in which each single gene controls the synthesis of a single polypeptide which may function independently or as a subunit of a more complex protein. (Replaces one-gene-one-enzyme hypothesis).

ontogeny, ontogenetic. The entire development of an organism (or part of it) from the zygote to maturity.

oo-. A prefix meaning egg.

oogamous, oogamy. Sexual reproduction between large, nonmotile eggs and smaller, motile sperm. Compare anisogamous and isogamous.

oogonium, *pl.* **oogonia.**   In certain algae and fungi, a unicellular gametangium that produces female gametes (eggs).

oosphere.   A large, nonmotile, female gamete; an egg.

oospore.   A thick-walled zygote characteristic of some fungi (Oomycetes) and algae.

open-dichotomous venation.   Dichotomous venation with free vein endings.

open venation.   Leaf venation in which large veins terminate blindly in the mesophyll instead of connecting with other veins.

operational taxonomic units (OTU).   Entities of the lowest rank used in any numerical taxonomic study. Should be representative of the organisms studied, e.g. individuals, lines, strains, species, etc.

operator.   A special chromosomal site at which the protein repressor product of the regulator gene acts. (*See* fig. O-2).

operculate.   Covered with an operculum.

operculum.   **1.** A lid, cap, or cover. In mosses, the lid on the capsule. (*See* fig. C-1); in certain fungi, a hinged cap on a sporangium or an ascus. **2.** The thickened tips of integuments of certain angiosperm ovules.

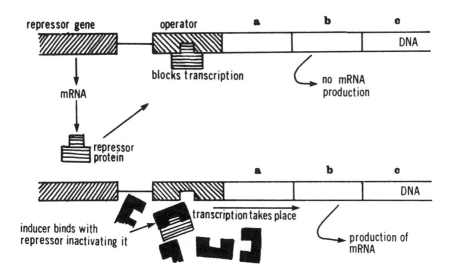

Fig. O-2. Operon: a,b,c. structural genes.

operon.   A group of genes consisting of an operator gene and the structural genes which it controls. (*See* fig. O-2).

opposite.   1. Having organs occurring in pairs on opposite sides of an axis, as with leaves or buds. **2.** With one organ in front of another, as stamens opposite petals.

orbicular, orbiculate.   More or less circular in outline or shape.

order.   A category of taxonomic classification below the class, composed of one or more families. The botanical suffix designating an ordinal name is: -ales.

ordinal.   Of or pertaining to an order.

organ.   A structure composed of different tissues organized to perform a specific function; one of the larger, visibly differentiated parts of a plant, such as root, stem or flower parts.

organelle.   Any of various structures within a cell with specialized structure and function, such as mitochondria, endoplasmic reticulum.

organ genera.   *See* form genera.

organic.   Of or pertaining to living things or to compounds formed by them.

organism.   Any individual living creature.

organogenesis.   The study of the development of organs.

orifice.   An opening.

ornamentals.   In horticulture, plants grown for their aesthetic qualities.

ornithophily, ornithophilous.   Pollination by birds.

ortet.   A plant from which material was taken for the purpose of vegetative propagation.

ortho-.   A prefix meaning straight or regular.

orthogenesis.   The tendency for a given development once started, to continue to an end due to mystical forces rather than natural selection, usually resulting in overspecialization. A discredited evolutionary theory.

orthostichy.   In phyllotaxy, a vertical line along which is attached a row of leaves or scales, on the axis of a shoot. Compare parastichy.

orthotropous ovule.   An erect ovule with the funicular attachment at one end and the micropyle at the other; i.e. one that is not bent or curved. *Syn.* atropous.

osmometer.   A device used to measure the rate of osmosis.

osmosis.   The diffusion of a solvent (usually water) through a differentially permeable membrane, from the side containing the higher concentration (of water), to the side containing the lower concentration. The migration continues until water concentration is equal on both sides.

osmotic potential.   *See* solute potential.

osmotic pressure.   The potential pressure that can be developed by a solution separated from pure water by a differentially permeable membrane. Serves as an index of solute concentration in a solution.

osseous.   Bony.

osteo-.   A prefix meaning bone.

osteosclereid.   A bone-shaped sclereid. (*See* fig. S-3).

ostiole.   An opening or pore; e.g. of a perithecium, pycnidium or conceptacle.

outbreeding.   The production of seed as a result of crossing genetically different plants (genotypes).

oval.   Broadly elliptical, the width more than half the length.

ovary.   The ovule-bearing region of a carpel in a simple pistil, or of a gynoecium composed of fused carpels (compound ovary). (*See* figs. F-3, M-9, O-3).

ovate.   Egg-shaped in outline, with the axis widest below the middle.

overdominance.   A situation in which the heterozygotes have a more extreme phenotype than either homozygote.

overtopping.   In the Telome theory, the dominance of one branch of a dichotomy over another; the dominant branch became a stemlike axis, the subordinate branch became "leaves."

ovoid.   A solid object that is oval in outline.

ovulate.   Bearing or possessing ovules.

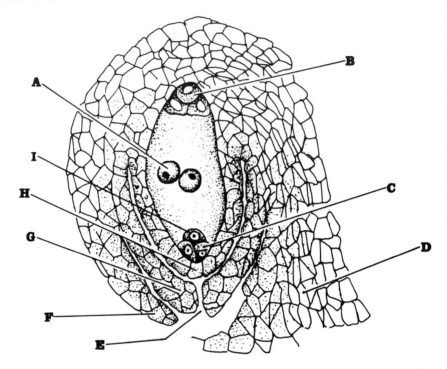

Fig. O-3. Ovule: A. polar nucleus, B. antipodals, C. synergids, D. funiculus, E. micropyle, F. outer integument, G. inner integument, H. nucellus, I. egg.

ovule.   The integumented megasporangium and enclosed structures which after fertilization become the seed. (*See* fig. I-4).

ovuliferous.   Ovule bearing, as in ovule-bearing structures.

ovuliferous scale.   A highly modified axillary branch that bears an ovule; in some conifers, a scalelike shoot in the female cone to which the ovule is attached.

ovum.   *See* egg.

oxidant.   A molecule that accepts the electron in an oxidation-reduction reaction, thus becoming reduced.

oxidation.   The loss of electrons from an atom or molecule; the addition of oxygen to a compound, or the loss of hydrogen.

oxidative phosphorylation.   The production of ATP from ADP and inorganic phosphate through oxidative processes; occurs in conjunction with the electron transport chain in mitochondria.

oxysomes. Stalked particles which occur on the internal membranes (cristae) of mitochondria; they occur in both plants and animals. *Syn.* elementary particle and inner membrane sphere.

P

P_1, P_2, P_3. Parental generations, e.g. first, second, etc. Used to designate parents when making crosses between plants, e.g. P_1 individuals would be the parents of an F_1 generation.

pachycauly. Having a short, thick, often succulent stem, as in certain species of the cactus family (Cactaceae).

pachymorph. One of two general types of rhizomes, characterized by a shortened, thick and fleshy stem, and determinate growth; its growth produces many-branched clumps which terminate in flowering stalks. Contrast leptomorph.

pachytene. The third stage of prophase I of meiosis, in which homologous chromosomes become tightly paired. *Syn.* pachynema and pachyphase.

paedomorphosis, paedogenesis. Having some characteristics that are juvenile and others that are mature in the same cell, tissue, organ or organism. *Syn.* neoteny.

palate. In sympetalous corollas, the raised projection of the lower lip which closes or very nearly closes the throat; e.g. in the figwort family (Scrophulariaceae).

palea. In grasses, the upper or inner of a pair of bracts (lemma and palea) that subtends the floret. (*See* fig. F-3).

paleaceous. Having small membranous scales; chaffy.

paleobotany. The study of the plant life of the geologic past.

paleozoic era. A geological era which began about 600 million years ago and lasted until about 280 million years ago. It was a time of active plant diversification and evolution. It includes, from oldest to youngest, the following geological time periods: Cambrian, Ordovician, Silurian, Devonian, Carboniferous and Permian.

palisade parenchyma. Chloroplast-bearing, leaf mesophyll cells which are elongated (columnar), with their long axes perpendicular to the surface of a leaf. *See* spongy parenchyma. (*See* fig. M-4).

palmate. Having lobes, veins (nerves) or divisions radiating from a common point, as in palmately lobed, palmately veined or palmately compound.

palmately compound. A compound leaf having leaflets radiating from a common point. *Syn.* digitate. (*See* fig. P-8).

250

palmatifid.   Cleft or divided palmately.

palmatisect.   Palmately divided into distinct segments.

palmella.   A nonmotile stage in the life history of some unicellular motile algae in which the daughter cells are surrounded by the gelatinous sheath of the parent cell; e.g. as in *Palmella,* a green alga.

palustrine.   A plant which grows in wet ground; a marsh inhabitant.

palynology.   The study of pollen grains and spores.

pandurate, panduriform.   Fiddle-shaped; more or less broadly obovate with a marked undulation along each side near the base.

panicle.   An indeterminate inflorescence, the main axis of which is branched, with pedicellate flowers borne upon the secondary branches. (*See* fig. P-1).

panicoid.   Pertaining to the grass subfamily, the Panicoideae.

paniculate.   Having a panicle type of inflorescence.

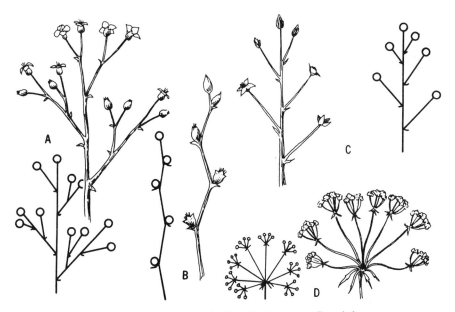

Fig. P-1. Panicle: A. panicle, B. spike, C. raceme, D. umbel.

panmictic unit.   A local population in which interbreeding is completely random, i.e. the individual members all have the same chance of mating and producing progeny. *Syn.* gamodeme.

panmixia.   Random mating, in which any individual is potentially capable of interbreeding with any other. *Syn.* panmixis.

pannose.   Feltlike; having a matted layer of trichomes with a very close texture; the appearance of felt.

pantoaperturate.   Pollen grains which have apertures scattered over the entire surface.

papain.   A protease enzyme found in the latex of the fruit and leaves of the papaya plant.

papilionaceous.   Descriptive of the "butterfly-like" flower of the legume subfamily Papilionoideae, which has a single large posterior petal (the banner or standard), two lateral petals (the wings), and usually two connate lower petals which form a single structure (the keel).

papilla, *pl.* **papillae.**   A soft, nipple-shaped protuberance; a type of trichome.

papillate, papillose.   Bearing papillae. *Syn.* shagreen.

pappose.   Having a pappus.

pappus.   A modified outer perianth-limb in the composites (Asteraceae) arising from the summit of the ovary and consisting of hairs, bristles, scales, awns or otherwise. Thought to be a modified calyx, but in some genera of composites, it may represent a modified corolla. (*See* fig. C-2).

para-.   A prefix meaning beside.

paracentric inversion.   A chromosomal inversion which occurs in only one arm of a chromosome and does not involve the centromere. Compare pericentric inversion. (*See* fig. P-2).

paracytic stoma.   A stomatal complex consisting of one or more subsidiary cells which flank the stoma parallel with the long axes of the guard cells. (*See* fig. S-16).

paradermal.   Refers to a section cut parallel to the surface (epidermis) of a flat organ such as a leaf.

parallel evolution.   *See* convergent evolution.

parallel venation, parallelodromous.   Having the primary veins in a leaf blade ar-

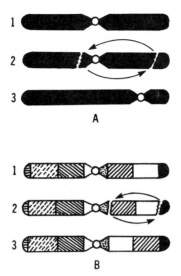

Fig. P-2. Paracentric inversion: A. pericentric inversion, B. paracentric inversion.

ranged approximately parallel to one another, although converging at the base and apex of the leaf; characteristic of many monocotyledons. *Syn.* striate venation.

parameter.   In statistics, the true value of a quantity for a population, as distinguished from estimates based on samples taken from the population.

paramutation.   A mutation in which one heterozygous allele permanently alters its allelic partner.

paramylum, paramylon.   A food reserve polysaccharide (β 1:3 linked glucan) characteristic of the euglenids (Euglenophyta).

paraphysis, *pl.* **paraphyses.**   Sterile hairlike, basally attached, filamentous structures occurring in the reproductive structures of many fungi (especially Ascomycetes), (*See* fig. A-17); also in bryophytes and brown algae (Phaeophyta).

paraphysoids.   Threads of hyphal tissue between asci; like paraphyses but without free ends.

parasexuality.   Somatic crossing-over in diploid nuclei which have arisen from the fusion of two genetically dissimilar nuclei in a heterokaryon. Eventually haploid nuclei are formed which show new combinations of genes thus deriving similar benefits of sexuality from a parasexual cycle. Characteristic of some fungi, as in *Aspergillus.*

parasite.   An organism that obtains nourishment from the living tissues of a host organism to the detriment of the host; frequently cause disease.

parasitism.   A type of symbiosis in which one symbiont benefits and the other is harmed.

parastichy.   In phyllotaxy, a helix along which is attached a series of leaves or scales on an axis. Compare orthostichy.

paratracheal parenchyma.   Axial parenchyma associated with vessels and other tracheary elements in the secondary xylem. Includes aliform, confluent and vasicentric.

paratype.   One of a group of specimens from which the description of a new species was prepared, other than the holotype or isotype.

parenchyma, parenchymatous.   Tissue composed of parenchyma cells.

parenchyma cell.   The most common cell type of plants, found in cortex tissue of stems and roots, in the pith of stems, in leaf mesophyll, in the flesh of fruits, and other areas; they are characteristically alive at maturity, usually thin-walled and exhibit a variety of sizes and shapes. Functions include photosynthesis, storage, secretion, movement of water and transport of food substances. (*See* fig. C-10).

parietal.   Located near or attached to the inner surface of a cell wall.

parietal cytoplasm.   Cytoplasm located next to the cell wall.

parietal placentation.   With ovules or placenta attached on the inner wall of simple or compound ovaries. (*See* fig. P-10).

paripinnately compound.   Even-pinnately compound, i. e. without a terminal leaflet.

parted.   Deeply cut more than half way to the middle or base of the structure.

parthenocarpy.   The natural or artificially-induced development of fruit without sexual fertilization. Such fruits are usually seedless, as in naval oranges, and commercially grown bananas.

parthenogenesis.   The development of an unfertilized egg into a new organism without sexual reproduction.

partial veil.   In some Basidiomycetes (club fungi), the hyphal tissue joining the margin of the cap to the stalk or stipe. The remains of the membrane left on the stalk after the expansion of the cap are called the annulus or ring. Also called the inner veil.

passage cell.   In certain roots, thin-walled endodermal cells with Casparian strips which occur among the thick-walled endodermis; they are situated opposite xylem cells, and most commonly found in roots of monocotyledons which undergo secondary growth.

passive trap.   A general term for structures of carnivorous plants which do not actively trap their prey, but instead depend on various methods of drowning them, or attracting them to sticky surfaces. Contrast active trap.

passive water absorption.   The entry of water into roots brought about by forces originating in the top of the plant, e.g. transpiration in leaves, and growth of stem tips, leaves and fruits.

pasteurization.   A method of destroying microbial populations by heating the liquid to a prescribed temperature for a specified period of time.

patelliform.   Knee- or disk-shaped.

patent.   Spreading.

pathogen.   An organism capable of causing disease or toxic symptoms.

pathology.   *See* plant pathology.

patristic similarity.   Similarity resulting from common ancestry.

pearl glands.   Short-stalked structures containing cells which are rich in highly nutritious contents, and serve as a food source for ants.

peat.   Partially decomposed plant tissue formed in water of marshes, bogs, or swamps, usually under conditions of high acidity. The three types of peat are moss peat, reed sedge peat and peat humus.

peat moss.   Partially decayed sphagnum moss used horticulturally as a soil conditioner, a propagating medium and for other uses; pH is about 3.5. Technically known as moss peat.

pectic substances.   A group of complex carbohydrates, derivatives of polygalacturonic acid. Occurs in plant cell walls and is an abundant constituent of the middle lamella. Occurs as protopectin, pectin (the basis of fruit jellies) and pectic acid.

pectin.   *See* pectic substances.

pectinate.   Comblike; having closely parallel, narrow, toothlike projections.

pedate.   Palmately lobed or divided with lateral lobes cleft or divided.

pedicel.   The stalk of an individual flower in an inflorescence or the stalk of a grass spikelet. Compare peduncle.

pedicellate.   Borne on a pedicel.

pedogenesis.   The development of soils, through the action of geochemical and biological processes.

peduncle.   The stalk of an inflorescence or the stalk of a solitary flower; the stalk bearing a strobilus. Compare pedicel.

pedunculate.   Borne on a peduncle.

peel method.   A method of preparing fossilized specimens for microscopic study. A surface of the fossil is smoothed, then etched with dilute acid and allowed to dry; a peel solution is poured over the etched surface, and allowed to dry; when this is peeled off, it contains a thin section of whatever plant materials were exposed from the etching.

pellicle.   Any surface, skinlike growth or covering as the network of interlocking flexible strips of protein around the protoplast in euglenids.

pellucid.   Nearly transparent in transmitted light; clear.

peltate.   Umbrella- or shield-shaped, with a stalk attached to the lower surface near the center of the structure; as in a peltate leaf or peltate hair. (*See* fig. P-3).

pendent, pendulous.   Suspended, nodding or hanging down from a support.

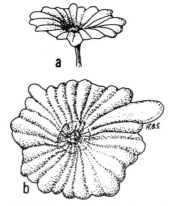

Fig. P-3. Peltate: a. side view, b. top view.

peneplain.   An erosional plain; a landscape of low relief formed by long-continued erosion.

penetrance.   The percentage of individuals that, when carrying a given gene in a combination that might allow its expression, actually express that phenotype; e.g. Mendel's genes showed 100 percent penetrance, but many others do not.

penicillate.   Ending in a tuft of fine hairs or branches.

penicillus, *pl.* **penicilli.**   The small, brushlike branches (conidiophore) which bear conidia in *Penicillium* (Fungi Imperfecti).

pennate diatoms.   A bilaterally symmetrical diatom usually of a rectangular shape, as in *Navicula*. (*See* fig. D-2).

penninerved.   Pinnately nerved.

penta-.   A prefix meaning five.

pentacyclic.   Having five whorls.

pentamerous.   Having parts in fives or multiples of five.

pepo.   A modified berry with a leathery nonseptate rind. Derived from an inferior ovary. Characteristic fruit of the pumpkin family (Cucurbitaceae).

peptide.   Two or more amino acids linked by peptide bonds.

peptide bond.   A covalent bond formed between two amino acids. The carboxyl end of one is joined to the amino end of the other; when this bond is made, a molecule of water is liberated.

percolation.   The downward movement of water through the soil.

perennating tissue.   Tissue, as in buds, which is capable of meristematic activity.

perennial.   A plant which lives for more than two years; herbaceous perennials have stems and/or leaves which are produced and die back annually, with their underground stems and/or roots remaining alive; woody perennials, e.g. trees and shrubs, have aerial stems which may live for many years.

perfect (flower).   A flower having both stamens (androecium) and carpels (gynoecium); may or may not have a perianth. *Syn.* monoclinous.

perfect stage.   The sexual phase in the life history of fungi that includes sexual fusion and the spores that result from such fusions.

perfoliate.   With the base of a leaf or petiole completely surrounding the stem.

perforation plate.   The perforated part of a wall of a vessel member; a plate may be simple, with only one perforation, or multiperforate.

peri-.   Prefix meaning around.

perianth.   **1.** A collective term for the floral envelopes, usually the combined calyx and corolla, or tepals of a flower. **2.** The modified organs surrounding a group of archegonia in liverworts.

periblem.   The meristem forming the cortex. One of the three histogens according to Hanstein.

pericarp.   **1.** The mature fruit wall which develops from the ovary wall; frequently differentiated into two or three distinct layers: exocarp, mesocarp, endocarp. (*See* figs. B-3, D-6). **2.** In the red algae (Rhodophyta), the urn-shaped covering of cells that surrounds the carposporophyte.

pericentric inversion.   A chromosomal inversion in which both chromosome arms are involved and thus includes the centromere. Compare paracentric inversion. (*See* fig. P-2).

perichaetium.   The leaves that surround or subtend the archegonia in some liverworts and many mosses.

periclinal.   **1.** Parallel to the surface. **2.** Refers to the cell wall orientation, or the plane of cell division. Contrast anticlinal. (*See* figs. A-23, P-4).

periclinal chimera.   A chimera in which cells of different genetic composition are arranged in periclinal layers.

pericycle.   A layer of tissue located inside the endodermis which forms a cylinder around the vascular tissues of many roots and stems of lower vascular plants; in seed plants, characteristic of roots but infrequent in stems. Branch roots arise from the pericycle. (*See* fig. M-2).

pericyclic fiber.   *See* perivascular fiber.

pericyclic sclerenchyma.   *See* perivascular sclerenchyma.

periderm.   The outer protective tissue that replaces the epidermis in those stems and roots which increase in thickness by secondary growth. Consists of phellem (cork), phellogen (cork cambium) and phelloderm. Commonly called bark.

peridiole.   The specialized glebal chamber which contains the basidiospores in the bird's nest fungi, Nidulariales (Basidiomycetes); often functions as an organ of spore dispersal.

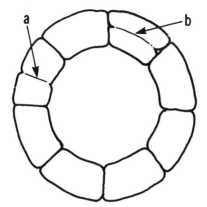

Fig. P-4. Periclinal: a. anticlinal division, b. periclinal division.

peridium.   The sterile outer covering or wall of a fungal fructification, as in the sporangium of slime molds (Myxomycetes) and certain club fungi (Basidiomycetes).

perigynium, *pl.* **perigynia. 1.** A saclike bract which surrounds the pistillate flower or the achene, e. g. in sedges of the genus *Carex.* **2.** In liverworts (Hepaticae), the sleevelike tubular extension of stem or thallus tissue which encloses the archegonia.

perigynous, perigyny.   Borne around the gynoecium. A floral condition in which the floral parts (calyx, corolla and stamens) are attached to the edge of a cup-shaped hypanthium. (*See* fig. E-4).

perimedullary region or zone.   In some stems, the outer, more or less morphologically distinct region surrounding the pith. *Syn.* medullary sheath.

periphysis, *pl.* **periphyses.**   Sterile, hairlike filaments lining the ostiole of a conceptacle, perithecium or pycnidium.

periplasm.   The outer nonfunctional protoplasm surrounding the egg (oosphere) of certain Oomycetes, e. g. Peronosporales.

periplasmodium. 1. The mass of tissue resulting from the disintegration and coalescence of tapetal cells. **2.** The multinucleate, mucilaginous mass derived from the sporangium wall of *Azolla.*

periplast.   A complex and often ornamented membrane surrounding the protoplasm of some algae that lack a cell wall; e.g. euglenids and cryptomonads (Cryptophyceae).

perisperm.   A food reserve tissue found in the seeds of certain plants which is derived from a diploid nucellus; e.g. coffee, beets, spinach. In comparison to endosperm, it is formed outside the embryo sac.

perispore. A wrinkled outer covering of some fern spores, e.g. ferns in the family Polypodiaceae.

peristome (teeth). In some mosses, multicellular tooth- or flaplike structures at the mouth (peristome) of the moss capsules. They facilitate spore dispersal by responding to changes in humidity. Under damp or humid conditions, the peristome teeth enclose the opening of the capsule preventing spore dispersal; when conditions become drier, the peristome folds back allowing release of spores.

perithecium. A globose or flask-shaped ascocarp in which the hymenium is completely enclosed except for a pore or ostiole at the top. (*See* fig. P-5).

peritrichous. A bacterial cell having flagella attached at many points.

perivascular fiber. A fiber located along the outer periphery of the vascular cylinder in the stem of a seed plant but outside the primary phloem, and therefore distinct from a primary phloem fiber. Also called a pericyclic fiber or a primary phloem fiber.

perivascular sclerenchyma. Sclerenchyma located along the outer periphery of the vascular cylinder and not originating in the primary phloem. Formerly called pericyclic sclerenchyma.

permafrost. In arctic and alpine regions, subsurface soil which remains permanently frozen throughout the year.

permanent wilting percentage. The amount of water remaining in the soil which cannot be removed by a particular plant growing in it; a plant in this condition is said to have reached its permanent wilting point.

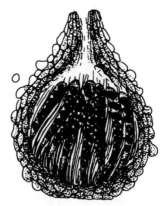

Fig. P-5. Perithecium.

peroxisome. A microbody which contains enzymes, and in which the glyoxylate cycle and photorespiration can occur; also believed to be involved in amino acid synthesis.

persistent. Remaining attached; not falling free.

personate. A bilabiate (two-lipped) corolla which has an arched upper lip and a lower lip that protrudes into and nearly closes the throat of the corolla.

perulate. Bearing scales, as most buds.

pesticide. A general term for any chemical used to kill or control plant pests, e.g. fungicide, herbicide, insecticide, etc.

pests (plant). A general term for all life forms which are destructive to plants, including diseases, predators and weeds.

petal. One of the members of the corolla of a flower. Frequently conspicuously colored.

petaloid. Resembling a petal in shape, texture and/or color.

petalostemonous. Having staminal filaments fused to the corolla and with the anthers free.

petiolar. Pertaining to a petiole.

petiolate. Having a petiole.

petiole. The stem of a leaf; the stalk attaching a leaf blade to a stem. (*See* fig. L-3).

petiolulate. Having a petiolule.

petiolule. The stalk of a leaflet of a compound leaf.

petri dish. A shallow, circular, transparent glass or plastic dish with an overlapping cover. Used for the culture of microorganisms, tissue culture, etc.

petrifaction. A type of plant fossil formed by the infiltration of cells and intercellular spaces by various carbonates, phosphates, silica, sulfides or other substances which subsequently become crystallized.

pH. A symbol denoting the relative concentration of hydrogen ions in a solution, that is, a measure of the acidity or alkalinity of a solution. pH values range from 0-14; pH 7 is neutral, less than 7 acidic, and greater than 7 alkaline.

phage. An abbreviated term for bacteriophage.

phagocyte.   A cell that ingests and destroys other cells, or foreign matter.

phagocytosis.   The intake of solid particles by a cell, by flowing over and engulfing them. Particles are attached to invaginations of the plasma membrane which subsequently enclose the particle in a vesicle, and are later digested. Compare pinocytosis.

phagotroph.   An organism that ingests solid food particles. *Syn.* macroconsumer.

phagotrophic, phagotrophy.   Ingesting solid food particles.

phalaenophily.   Pollination by moths.

phanerocotylar.   Having cotyledons which emerge from the seed coat; cotyledons usually appearing above ground. Contrast cryptocotylar.

phanerogam.   A plant that reproduces by seeds; a spermatophyte. Compare cryptogam.

phanerophyte.   Woody plants or herbaceous evergreen perennials that grow taller than .5 meters, or whose shoots do not die back periodically to .5 meters. Compare chamaephyte.

phaneroplasmodium, *pl.* **phaneroplasmodia.**   A plasmodium consisting of a well-differentiated advancing fan and conspicuous thick strands in which ecto- and endoplasmic regions are well differentiated, and in which the protoplasm is coarsely granular; characteristic of the Myxomycetes (slime molds).

phellem.   *See* cork. (*See* fig. P-6).

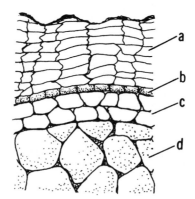

Fig. P-6. Phelloderm: a. phellem, b. phellogen, c. phelloderm, d. cortex.

phelloderm. A tissue produced inwardly by the cork cambium, as cork is produced toward the outer surface. The phelloderm forms the innerpart of the periderm. (*See* fig. P-6).

phellogen. *See* cork cambium. (*See* fig. P-6).

phelloid cell. A nonsuberized cell within the phellem (cork) but resembling cork cells; they occur in various proportions and arrangements within the phellum and may be sclerified in some plants.

phenetic. A classification system based on current total simularity, using as many separate characteristics as possible, each of which is given equal weight; e. g. morphological, physiological, anatomical or biochemical criteria. Such a system does not reflect phylogeny because the evolutionary history is unknown.

phenogram. A graph illustrating phenotypic relationships.

phenology. The study of periodicity in organisms as related to climatic events; in plants, the various biological processes that are correlated with the seasons, e. g. flowering, fruiting, dormancy, root or stem elongation, etc.

phenotype. The external physical appearance of an organism resulting from interaction between the genetic constitution (genotype) of the individual and the environment.

phenotypic plasticity. *See* plastic.

pheromones. Chemical substances secreted by an animal which influence the behavior and/or morphological development of other animals of the same species; pheromones may also be used as a means of communication between organisms.

phialide. A small bottle-shaped structure from which spores (conidia) are produced, as in certain Ascomycetes, e.g. *Penicillium;* conidia may be formed inside the phialide and extruded, or they may bud off the phialide.

phialospore. A spore produced from a phialide.

-phile, -philous. A suffic meaning loving.

phlobaphenes. Anhydrous derivatives of tannins, which are very conspicuous in sectioned plant material; appear as yellow, red or brown, amorphous, granular masses of various sizes.

phloem. The principle food-conducting tissue of a vascular plant which is usually composed of sieve elements, various kinds of parenchyma cells, fibers and sclereids. (*See* figs. A-23, C-4, C-8, C-12, M-4).

phloem initial.   A cambial cell that divides periclinally one or more times and gives rise to cells that differentiate into phloem elements. *Syn.* phloem mother cell.

phloem parenchyma.   Parenchyma cells located in the phloem. In secondary phloem, it refers to axial parenchyma.

phloem ray.   A portion of a vascular ray located in the secondary phloem. (*See* fig. C-12).

phloic procambium.   That portion of procambium which differentiates into primary phloem.

-phore.   A suffix meaning carrying, supporting or bearing.

phosphate.   A compound of phosphorus; in general, phosphoric acid.

phospholipid.   A lipid molecule to which a phosphate group is attached. They are important components of cell membranes.

phosphorylation.   A chemical reaction in which phosphate is added to a compound; e.g. the formation of ATP from ADP and inorganic phosphate.

photic zone.   In an aquatic environment, the zone penetrated by light.

photo-, -photic.   A prefix or suffix meaning light.

photoautotroph, photoautotrophic.   Organisms, like green photosynthetic plants, that manufacture their own food from inorganic compounds using light energy.

photobiology.   The study of the biological effects of the cyclicity, the spectral quality, and the intensity of light on plants and animals.

photolysis.   Splitting of water into H^+ and OH^-, utilizing solar energy in the light reactions of photosynthesis.

photomorphogenesis.   Development which is influenced (either promoted or inhibited) by light, and is independent of the direction of illumination.

photonastic, photonasty.   Response of plants to changes in illumination, e. g. opening and closing of flowers.

photoperiod.   Day length; often used specifically as the length of exposure to light required for some developmental functioning of an organism.

photoperiodism.   The response of plants to the relative duration (length) of day and night; influences many activities of plants including growth, fruit development, flowering, and preparation for winter dormancy. *Syn.* photoperiodicity.

photophosphorylation. The production of ATP by a phosphorylation reaction using the energy fixed in the light reactions of photosynthesis.

photoreceptor. A structure containing a light-sensitive pigment.

photorespiration. The light-dependent production of glycolic acid in chloroplasts and its subsequent oxidation in peroxisomes. This process, which occurs primarily in C_3 (Calvin cycle) plants, greatly decreases plant productivity because large amounts of certain early products of photosynthesis are respired away.

photosynthesis. The process whereby light energy is converted into chemical energy in the presence of chlorophyll. It involves the production of a carbohydrate from carbon dioxide and water with the release of oxygen.

phototaxis. The movement or response of an organism or cell to the stimulus of light.

phototroph, phototrophic. An organism that derives its nutrition from photosynthesis.

phototropism. A growth movement which is influenced by the direction of light; bending or turning of an organism in response to light, as growth of a plant in the direction of a light source. Movement may be positive (toward the light) or negative (away from light).

phragmoplast. A spindle-shaped system of fibrils (composed of microtubules) that appears at telophase of mitosis between the two daughter nuclei. It plays a role in the formation of the cell plate during cell division (cytokinesis).

phragmosome. A cytoplasmic layer formed across the cell prior to the formation and in the same plane as the phragmoplast; it forms a medium in which the phragmoplast and cell plate develop. In electron microscopy, it refers to the microbodies that are thought to participate in cell plate formation.

phreatophyte. A perennial plant which is very deep rooted, deriving its water from a more or less permanent, subsurface water supply; it is thus not dependent upon annual rainfall for survival.

phycobilins. A group of water-soluble, protein-linked pigments found in blue-greens (Cyanophyta), red algae (Rhodophyta) and the cryptophyceans (Cryptophyaceae).

phycobilosomes. Minute particles or bodies containing phycobilins which occur on the surface of the thylakoids of blue-greens, red and cryptophycean algae.

phycobiont. The algal component of a lichen.

phycocyanin. A blue phycobilin pigment found in blue-greens, red and cryptophycean algae.

phycoerythrin. A red phycobilin pigment found in blue-greens, red and cryptophycean algae.

phycology. The study of algae. *Syn.* algology.

phycomycete. Any individual fungal member of the class Phycomycetes.

phylad. An evolutionary line; a natural group of common ancestry.

phyletic. *See* phylogenetic.

phyletic slide. In ferns, an apparent evolutionary progression of the sporangia from a marginal to an abaxial (lower surface) position on the leaf.

phyllary. An individual bract of the involucre found in the composites (Asteraceae). (*See* fig. C-2).

phyllo-, phyll-, -phyll. A prefix or suffix meaning leaf.

phylloclade. *See* cladode.

phyllode. *See* phyllodium. (*See* fig. P-7).

phyllodium, *pl.* **phyllodia.** An expanded, leaflike petiole but without a true blade, which functions in photosynthesis; e.g. as in some acacias and other plants. *Syn.* phyllode. (*See* fig. P-7).

Fig. P-7. Phyllode.

phyllome. A generalized term for the foliar organs of a plant, i.e. leaves, bracts, scales and floral appendages.

phyllopodic. Having well-developed, blade-bearing leaves at the base of a plant.

phyllosporous. With the ovules borne on specialized or modified leaves, rather than on the ends of telomes.

phyllotaxis, phyllotaxy. The arrangement of leaves along a shoot axis, e.g. opposite, whorled, decussate, etc. When expressed numerically as a fraction: the numerator represents the number of revolutions of a spiral made when going from one leaf around the axis until the leaf directly above the initial leaf is reached and the denominator represents the number of leaves passed in the spiral made as described above.

phylogenetic, phylogeny. Pertaining to the evolutionary history and relationships among a group of organisms.

phylogeny. The evolutionary history and relationships among a group of organisms.

phylum, *pl.* **phyla.** *See* division.

physiognomy. The external appearance, or morphology of the vegetation of a plant community; also refers to the overall composition of a community, e. g. forest, grassland, desert, etc.

physiological races. Subdivisions of a species or variety identical in morphology but differing in certain physiological, biochemical, or pathological characters; e. g. pathogens of the same species which differ in their ability to parasitize a given host; includes most species of plant pathogenic fungi.

physiology. The study of life activities and metabolic processes of organisms.

phyto-, -phyte. Prefix or suffix meaning plant.

phytobenthon. A collective term for plants attached to the bottom of aquatic habitats.

phytochrome. A phycobilin-like pigment found in the cytoplasm of green plants. It acts as a photoreceptor for red/far-red light and is involved in a number of timing processes, e.g. flowering, dormancy and seed germination.

phytocoenosis. A plant community.

phytogeography. The study of the distribution of plants.

phytohormone. A plant hormone.

phytological.   Relating to the study of plants.

phytoplankton.   A collective term for free-floating, or weakly swimming aquatic plants, e. g. certain algae and diatoms.

phytotoxins.   Chemicals produced by plants which are toxic to some plants and animals.

pigment.   A molecule that absorbs light, usually selectively, (i.e. only certain wavelengths), e.g. chlorophyll.

pileate.   Having a cap or caplike structure.

pileus, *pl.* **pilei.**   The cap or caplike structure of a mushroom which bears the hymenium on the lower surface. Characteristic of some sac (Ascomycetes) and club (Basidiomycetes) fungi. (*See* fig. B-2).

piliferous.   Hairlike, flexuous, many times longer than wide.

pillar cell.   A subepidermal sclereid in the seed coat of certain legumes (Fabaceae). *Syn.* osteosclereid and hourglass cell.

pilose.   Having soft, long, shaggy trichomes.

pilosism.   Abnormal hairiness in plants.

pilosulous.   Diminutive of pilose.

pin flower.   In heterostylous plants, the flower type having the longer style. Compare thrum flower. (*See* fig. D-5).

pinna, *pl.* **pinnae.**   A primary subdivision, or leaflet, of a compound leaf or frond. Pinnae may be further subdivided into pinnules.

pinnate.   Shaped like a feather; e. g. having leaflets of a compound leaf arranged on opposite sides of a common axis or rachis.

pinnately compound.   A compound leaf with leaflets arranged pinnately along a common axis, the rachis. (*See* fig. P-8).

pinnately netted or veined.   Having a pinnate vein pattern.

pinnate-pinnatifid.   Pinnate with pinnatifid pinnae.

pinnatifid.   Pinnately cleft or divided.

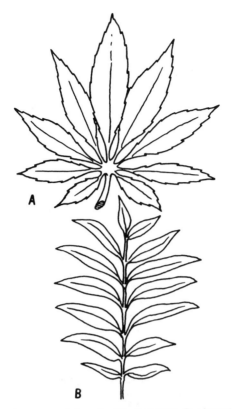

Fig. P-8. Pinnately compound: A. palmately compound, B. pinnately compound.

pinnatisect. Pinnately cut to the midrib into distinct segments.

pinninerved. Pinnately veined.

pinnule. A secondary subdivision of a pinna of a pinnate leaf. (*See* fig. A-9).

pinocytosis. The intake of liquid particles by a cell, e.g. mineral nutrients and various macromolecules. Compare phagocytosis.

piriform. Shaped like a pear.

pisiform. Pea-shaped.

pistil. The female reproductive organ of a flower composed of an ovary, style and stigma. A pistil may consist of a single carpel or it may consist of two or more fused carpels.

pistillate.   Pertaining to a flower having one or more carpels but no functional stamens; unisexual and female. (*See* figs. C-15, S-10).

pistillode.   A rudimentary or vestigial pistil present in some staminate flowers.

pit.   A recess or cavity in a cell wall where the primary cell wall is not covered by a secondary cell wall. *See* bordered and simple pit. (*See* fig. P-9).

pit aperture.   The opening into the pit from the interior of the cell.

pit canal.   The passage from the cell lumen to the chamber of a bordered pit.

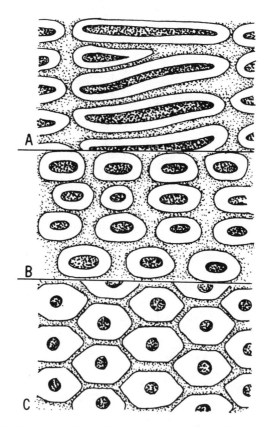

Fig. P-9. Pits: A. scalariform pitting, B. opposite pitting, C. alternate pitting.

pit cavity, pit chamber.   The entire area within a pit from the pit membrane to the cell lumen or to the outer pit aperture if a pit canal is present. (*See* fig. B-6).

pitcher.   An inflated, tubular-shaped insectivorous leaf, as in the pitcher plant, *Sarracenia.*

pitcher plants.   One of many insectivorous plants which trap their prey by means of passive traps, e.g. plants in the genus *Sarracenia.*

pit-field.   *See* primary pit-field.

pith.   The ground tissue which occupies the center of a stem or root inside the vascular cylinder; it usually consists of parenchyma cells.

pith ray.   *See* interfascicular region.

pit membrane.   *See* margo. (*See* fig. B-6).

pit-pair.   A pair of complementary pits of two adjacent cells. Consists of two pit cavities and a pit membrane.

pitted.   Having little depressions or cavities.

placenta, *pl.* **placentae.**   The portion of the ovary wall which bears the ovules or seeds. (*See* figs. B-3, S-7).

placentation.   The arrangement and distribution of placentae and ovules within the ovary. *See* axile, basal, free-central and parietal placentation. (*See* fig. P-10).

plagiotrophic growth.   Horizontal growth of plant organs, e. g. roots, branches, rhizomes.

plain-sawed wood.   Lumber sawed so that the tangential surface is exposed on the surface of the board.

planation.   An evolutionary process in which a three-dimensional branch system becomes flattened into a single plane, as may have occurred during the evolution of the megaphyll (leaf).

plane.   A flat, level or even surface.

plankton.   The free-floating or weakly swimming aquatic microorganisms; includes phytoplankton (e. g. certain algae and diatoms) and zooplankton.

planogamete.   A motile gamete.

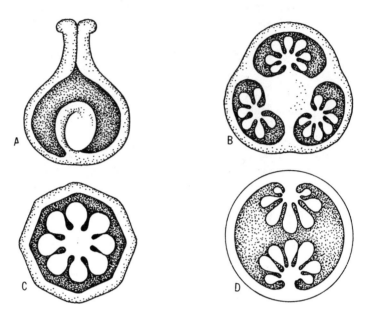

Fig. P-10. Placentation: A. basal, B. axile, C. free central, D. parietal.

planogametic copulation.   The fusion of naked gametes, one or both of which are motile, e.g. some chytrid fungi.

planospore.   A motile asexual spore; also called a zoospore.

plant hormone.   *See* hormone.

plantlet.   A small, immature plant.

plant pathology.   The study of plant diseases.

plaque.   A round, clear area in a bacterial culture which is the result of bacterial lysis by viral (bacteria phage) multiplication.

-plasma, plasmo-, -plast.   Prefix or suffix meaning formed or molded.

plasma membrane, plasmalemma.   The outer three-ply membrane bounding the protoplast next to the cell wall. Also called cell membrane or ectoplast.

plasmodesma, *pl.* **plasmodesmata.**   Minute cytoplasmic connections that extend through pores in the cell walls between protoplasts of adjacent living cells. They are thought to provide cytoplasmic continuity between adjacent cells.

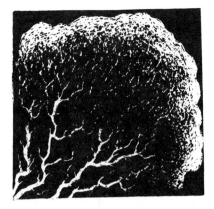

Fig. P-11. Plasmodium of a slime mold.

plasmodiocarp. A slime mold (Myxomycete) sporocarp which develops from strands of plasmodia. It resembles a sessile, elongate, branched sporangium.

plasmodium, *pl.* **plasmodia. 1.** A membrane bounded, multinucleate, amoeboid, mass of protoplasm. **2.** The vegetative phase in the life cycle of slime molds (Myxomycetes). (*See* fig. P-11).

plasmogamy. Union or fusion of the protoplasts of two (haploid) gametes, without the fusion of their nuclei; characteristic of Ascomycetes and Basidiomycetes.

plasmolysis. The separation of the cytoplasm from the cell wall caused by the removal of water from the protoplast. Occurs when the concentration of water in a solution outside a cell becomes less than that inside the cell, resulting in an outward movement of water by osmosis.

plastic, plasticity. 1. The ability or degree to which a given genotype may vary phenotypically under different environments. **2.** The total variation (both genotypic and phenotypic) of a species or population. *Syn.* phenotypic plasticity.

plastid. In certain eukaryotes, a double membrane bounded cytoplasmic organelle that functions in such activities as photosynthesis and storage.

plastochrone. The time interval between the inception of two successive repetitive events; i. e. the time interval between various corresponding stages in the development of leaves, internodes, axillary buds, floral organs, etc. Also spelled plastochron.

plate meristem. A meristematic tissue consisting of parallel layers of cells which divide only anticlinally with reference to the wide surface of the tissue. Characteristic of plant parts that assume a flat form, as a leaf.

platycanthous.   Having flat and usually large spines.

playa.   A depression in an arid or semiarid region that temporarily fills with runoff water after rainstorms.

plectenchyma.   A general term used to designate all types of fungal tissues. The two most common types are prosenchyma and pseudoparenchyma.

plectostele.   A protostele in which bands of xylem alternate with the phloem (as viewed in cross section); e.g. *Lycopodium.*

pleio-, pleo-.   A prefix meaning several, full or more.

pleiotrophy, pleiotropism.   The capacity of a single gene to affect a number of different characteristics.

pleomorphic, pleomorphism.   **1.** Having more than one shape or form; polymorphic. **2.** Actinomorphic with a reduction in the number of parts.

plerome.   According to Hanstein, the meristem forming the core of the root or stem which is composed of the primary vascular tissues and associated ground tissue. One of the three histogens, plerome, periblem and dermatogen.

pleur-.   A prefix meaning to a side.

pleurocarpous.   A growth form of mosses in which the gametophyte is much branched and creeping, with the sporophytes borne at the tips of short lateral branches. Compare acrocarpic.

pleurogamy.   The condition wherein an ovule is penetrated laterally through its side by a pollen tube, prior to fertilization.

pleuropneumonia.   *See* mycoplasmas.

plexus.   A network of interconnecting linear structures, as veins.

plicate.   Plaited; having a series of longitudinal folds, like a fan.

pliestesial.   Monocarpic, but living several years before flowering a single time; as in the century plant, *Agave.*

-ploid.   A suffix used to designate a particular multiple of the chromosome set in the nucleus of an organism, e.g. 12-ploid, 24-ploid, etc.

ploidy, ploidy level.   Referring to the number of chromosomes in a cell, tissue, or organism. *See* polyploid.

plumose. Covered with a fine, featherlike pubescence.

plumule. The first bud of an embryo, the part of the embryonic axis above the cotyledonary node.

plur-. A prefix meaning several or many.

plurilocular. **1.** Having many locules or cells. **2.** A condition of sporangia or gametangia wherein they contain numerous cells, each of which is fertile and produces a reproductive cell or spore. (*See* fig. P-12).

pneumatocyst. A hollow, air-containing bulb at the base of the leafy blades of some brown algae (Phaeophyta), which help in keeping the plant afloat.

pneumatophores. Vertical extensions of the buried roots of certain trees (e.g. man-

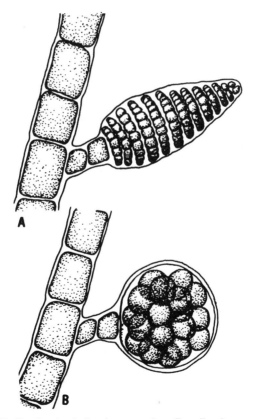

Fig. P-12. Plurilocular: A. plurilocular sporangium, B. unilocular sporangium.

groves) which exist in marsh or swamp habitats. They extend out of the mud or water and function as pathways for the exchange of gases between the atmosphere and the submerged roots. *Syn.* air roots.

pod.   A general term applied to any dry, dehiscent fruit.

podetium, *pl.* **podetia.**   An erect, stalklike, cup- or funnel-shaped structure which arises from the thallus of certain lichens; e.g. *Cladonia.*

poisson distribution.   A distribution in which the objects in space or the events in time are completely random and in no way influence the occurrence of any other of the objects or events in time or space.

polarity.   The condition exhibited by living cells or organisms in which one end is differentiated from the opposite or contrasting end, by structure and/or function. For example, stems and roots of many plants frequently exhibit a marked degree of polarity making it difficult or impossible to root cuttings of these plants unless their correct polarity is maintained, i.e. with the apical end of a stem cutting "up", and with the end toward the root, "down"; usually the reverse is true for roots. Leaves show the least amount of polarity.

polar nodule.   A thickening in the wall at each end of the groove or raphe of certain diatoms and in the heterocysts of blue-greens.

polar nuclei.   In the embryo sac of angiosperms, the mitotically formed nuclei (usually two), which migrate from the poles to become centrally located. They fuse with a sperm upon fertilization, and form the primary endosperm nucleus. (*See* figs. E-2, O-3).

pollard.   A tree which has had its crown removed to promote growth of a thick mass of shoots from the point of cutting.

pollen.   A collective term for the pollen grains or microspores of seed plants.

pollen chamber.   A chamber at the top of the megasporangium of gymnosperms in which pollen collects. It is open to the outside through the micropyle. *See* pollination drop.

pollen flower.   A flower which produces pollen but no nectar.

pollen grain.   A microspore of a seed plant which contains a mature or immature microgametophyte (male gametophyte). (*See* figs. I-4, P-14, P-17, T-1).

pollen mother cell.   *See* microsporocyte.

pollen sac.   A cavity in the anther which contains the pollen grains.

pollen tube.   A tubular, unicellular extension formed following germination of the pollen grain. It carries the male gametes (sperm) through the stigma and style into the ovule. (*See* figs. I-4, P-14, P-17).

pollinate, pollination.   The transfer of pollen from an anther to a stigma in angiosperms; or from a microsporangium to the micropyle of the ovule in gymnosperms.

pollination drop.   A droplet of fluid secreted through the micropyle of gymnosperms at the time of pollination. It traps airborne pollen grains and as it evaporates, it draws the pollen into the pollen chamber where it contacts the surface of the nucellus.

polliniferous.   Bearing pollen.

pollinium, *pl.* **pollinia.**   A cohesive mass of pollen grains which are shed together and transported as a unit during pollination, as in the orchids (Orchidaceae) and milkweeds (Asclepiadaceae). (*See* fig. P-13).

poly-.   A prefix meaning many.

polyad.   Compound pollen grains consisting of clusters of more than four grains.

polyandrous, polyandry.   Having many stamens.

polyarch.   Having many protoxylem strands or poles; as in the primary xylem of certain roots.

polycarpous.   With many separate carpels.

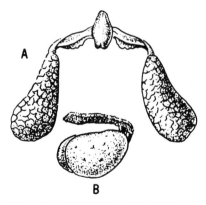

Fig. P-13. Pollinium: A. milkweed, B. orchid.

polycentric (thallus).   The growth form of certain chytrid fungi (Chytridiales) in which many sporangia are formed. Contrast monocentric.

polycentric chromosome.   Chromosomes which contain diffuse centromeres.

polycephalous.   Bearing many heads or capitula, as in composites (Asteraceae).

polydelphous.   Having many groups of stamens which are connate by their filaments.

polyderm.   A modified periderm which occurs in the roots and underground stems of several dicot families. This tissue contains alternating layers of living suberized and nonsuberized parenchyma cells.

polyembryony.   The condition of having more than one embryo in an ovule.

polygamo-dioecious.   Plants functionally dioecious but having a few perfect flowers on otherwise staminate or pistillate plants.

polygamo-monoecious.   Plants functionally monoecious but having a few perfect flowers.

polygamous.   Plants bearing both perfect and imperfect flowers.

polygenes.   A set of genes that function together to control a quantitative characteristic. Individually, the effects of the genes are too slight to identify. *Syn.* redundant DNA.

polyhaploid.   An organism in which the gametic chromosome number arises as a result of parthenogenesis in a polyploid.

polylectic.   *See* polytropic.

polymer.   A large molecule composed of many molecular subunits that are chemically identical.

polymerization.   The chemical union of monomers such as glucose or nucleotides to form polymers such as starch or nucleic acid.

polymerous.   Having many members in each cycle or series; as a floral whorl with many members.

polymery.   The situation in a multiple gene system in which the individual genes have equal and additive effects.

polymorphic, polymorphous.   Having several to many variable forms within the same species.

polymorphism.   The occurrence of two or more genetically distinct variants of a single species in the same interbreeding population at frequencies greater than can be explained by recurrent mutation.

polynucleotides.   Long-chain molecules composed of monomers (units) called nucleotides; e.g. DNA.

polypeptide.   A chain of many amino acids linked together by peptide bonds; the molecule formed by the union of numerous amino acids.

polypetalous, polypetaly.   A corolla of separate and unfused petals.

polyphenism.   Having several phenotypes in a single population which are not due to genetic differences between the individuals.

polyphily.   A flowering plant which is pollinated by many different taxa of vectors.

polyphyletic.   The origin of organisms from more than one ancestor. Compare monophyletic.

polyploid, polyploidy.   A plant, tissue, or cell having more than two complete sets of homologous chromosomes.

polyribosome.   *See* polysome.

polysaccharide.   A carbohydrate composed of many monosaccharide monomers (units) joined in a long chain, e. g. starch, cellulose and glycogen.

polysepalous, polysepaly.   A calyx of separate and unfused sepals.

polysiphonous pollen grains.   Pollen grains which form more than one pollen tube; multiple tubes may develop on the sitgma upon germination, or branches may develop from a single tube, before or after fertilization. However, only one pollen tube usually develops enough to effect fertilization. Compare monosiphonous. (*See* fig. P-14).

polysome, polyribosome.   A row of ribosomes situated along a strand of messenger RNA; the site of protein synthesis.

polysomic.   A normal diploid chromosome set which has one or more extra chromosomes, e. g. 2N + 1, trisomic or 2N + 2 tetrasomic.

polyspermy.   The penetration of the egg by more than one sperm at the time of fertilization.

polystelic.   Containing more than one stele.

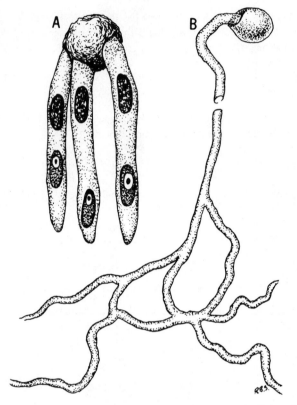

Fig. P-14. Polysiphonous pollen grain: A. many pollen tubes from one grain, B. branched pollen tube.

polystichous.   Having many vertical ranks or rows, as in leaves.

polytopic.   The occurrence of a taxon in two or more separate areas.

polytropic.   An animal vector which collects food, primarily pollen and nectar, from several taxa of flowering plants. *Syn.* polylectic.

pome.   A simple fleshy fruit, the outer portion being formed by the expanded floral parts and receptacle. Characteristic of only one subfamily of the rose family (Rosaceae), e. g. apples and pears.

population.   Any group of organisms of the same species living in a particular area at the same time. *Syn.* deme.

population genetics.   The study of the genetic composition of populations.

porate. Having pores; a pollen grain with one or more pori.

pore. 1. A small more or less round aperture. **2.** A vessel in the secondary xylem as viewed in cross section.

pore multiple. In secondary xylem, a group of two or more pores crowded together and flattened along the surfaces of contact.

poricidal. Opening or dehiscing by pores.

porogamy. The condition wherein an ovule is penetrated through the micropyle by a pollen tube, prior to fertilization. Compare chalazogamy and mesogamy.

pororate. A pollen grain with one or more compound pori.

porous wood. Secondary xylem which has vessels.

porrect. Directed outward and forward; perpendicular to the surface.

porus, *pl.* **pori.** Circular or faintly elliptic apertures on the surface of pollen grains.

postemergent spray. A pesticide or herbicide spray that is applied after a crop has emerged from the soil.

posterior. In plants, the side toward the axis; the upper side.

ppm. An abbreviation meaning parts per million, as in the amount of a substance in a solution or mixture.

P-protein, phloem protein. A proteinaceous substance found in phloem cells of angiosperms, most commonly in the sieve elements. Also called slime.

praemorse. Terminating abruptly, as if bitten off; said of roots.

precocious. Developing very early; often said of plants that flower before the appearance of the leaves.

precursory cell. A cell which gives rise to others by division. *Syn.* mother cell.

preemergent spray. A pesticide or herbicide spray that is applied after planting, but before the crop emerges from the soil.

preprophase band. A ringlike band of microtubules that delimits the equatorial plane of the future mitotic spindle in many cells.

prickle. A sharp-pointed epidermal or cortical outgrowth.

prickly pear.   A cactus of the genus *Opuntia,* so named because of its glochid covered fruits.

prim-.   A prefix meaning first.

primary cell wall.   The original cell wall layer of a plant cell which is deposited during the period of cell expansion.

primary endosperm nucleus.   The nucleus formed as a result of the fusion of a sperm nucleus and the polar nuclei (usually two). Gives rise to the endosperm.

primary growth.   Growth resulting from the production of cells by the apical meristems (protoderm, procambium and ground meristem). Results in an increase in length. Compare secondary growth.

primary hypha.   A uninucleate, haploid hypha produced from an ascospore or a basidiospore.

primary infection.   The first infection caused by a pathogen after going through a resting or dormant period.

primary meristem.   The three kinds of meristematic tissues derived from the apical meristem, e.g. protoderm, procambium and ground meristem.

primary phloem.   Phloem tissue derived from the procambium during primary growth. The first primary phloem formed is termed protophloem, later formed phloem is termed metaphloem.

primary phloem fiber.   *See* perivascular fiber.

primary pit.   *See* primary pit-field.

primary pit-field.   A thin area in a primary cell wall and middle lamella through which plasmodesmata pass. Also called primary pit and primordial pit.

primary plant body.   That part of the plant body arising from the apical meristems and their derivative meristematic tissues; it is composed entirely of primary tissues.

primary root.   The first root of the plant which develops as a continuation of the root tip (radicle) of the embryo. Becomes the tap root in gymnosperms and dicots, and soon dies in monocot seedlings, being replaced by adventitious roots.

primary succession.   Ecological succession that begins on bare land that has not previously been occupied by organisms.

primary suspensor.   The tier of cells above the embryo-forming cells in some conifers.

primary thickening meristem.   A meristem derived from the apical meristem and responsible for the primary increase in thickness of the shoot axis. Often found in monocotyledons.

primary tissue.   Cells derived from the primary meristematic tissues of the root and shoot. Compare secondary tissue.

primary vascular tissue.   The vascular tissues differentiated from the procambium.

primary wall.   *See* primary cell wall.

primary xylem.   Xylem tissue derived from the procambium during primary growth. The first primary xylem formed is termed protoxylem, later formed xylem is termed metaxylem.

primitive.   The ancestral condition; being little changed with time or through evolution.

primocane.   The first year's shoot or cane which normally does not flower, e. g. as in most brambles. *Syn.* turion.

primordial pit.   *See* primary pit-field.

primordium, *pl.* **primordia.**   A cell, tissue, or organ in its earliest stage of differentiation, e. g. leaf primordium. *Syn.* anlage.

prison flower.   *See* trap blossom.

pro-.   A prefix meaning before, in front of, forward.

proanthesis.   Flowering before the normal period, as spring flowers appearing in the fall.

probability.   The ratio of the proportion of a specified event to the total number of events.

probasidium, *pl.* **probasidia.**   An immature basidium in which karyogamy and meiosis normally occur.

procambium.   One of the three primary meristematic tissues. Procambium gives rise to the primary vascular tissues, xylem and phloem, and vascular cambium (if present). *Syn.* provascular tissue. (*See* figs. M-2, P-15).

prochromosomes.   Small condensed, heterochromatic (dark staining) segments of entire chromosomes which are found in interphase nuclei of many plant species. The number of prochromosomes may or may not correspond to the actual diploid chromosome number.

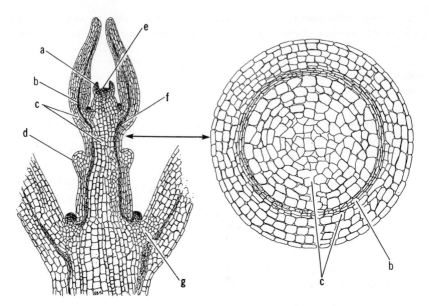

Fig. P-15. Procambium: a. leaf primordium, b. procambium, c. ground meristem, d. node, e. apical meristem, f. leaf trace, g. leaf gap.

procumbent.   Trailing or lying on the ground without rooting at the nodes. *Syn.* prostrate.

proembryo.   An embryo in the early stages of development, prior to the differentiation of the suspensor and the main embryo body.

progametangium, *pl.* **progametangia.**   A cell or group of cells which develops into the gametangium.

progeny test.   The evaluation of a parental genotype by studying its progeny under controlled conditions.

prokaryotic, procaryotic.   Organisms which lack a nucleus and other membrane-bound organelles, and lack chromosomes, plastids, and Golgi apparati. Examples include blue-greens, bacteria and mycoplasms. Contrast eukaryotic.

prolamellar body.   A poorly developed chloroplast characteristically found in meristematic cells in the root or shoot and in dark-grown plants. It contains few lamellae which form through the fusion of vesicles produced by invaginations of the inner layer of the double membrane.

promeristem, protomeristem. The apical meristematic cells and their most recent derivatives in the shoot or root.

promoter. A specific site between the operator and the first structural gene in an operon at which the formation of the mRNA begins.

promycelium, *pl.* **promycelia.** The basidium of the rusts and smuts which develops from the teliospore.

propagation. The reproduction of a plant by sexual or vegetative (asexual) means.

propagule. Any structure, sexual or vegetative (asexual), that becomes separated from the parental individual and serves as a means of propagation for that plant. *Syn.* diaspore.

prophage. A noninfectious phage which is linked with the bacterial DNA and replicates with the bacterial DNA, but does not bring about lysis of the bacteria.

prophase. An early stage of mitosis in which the nuclear envelope breaks down, the chromosomes become visible and begin to migrate toward the metaphase plate, the nucleolus disappears and the spindle forms.

prophase I. An early stage in meiosis in which the homologous chromosomes pair, chiasmata form and crossing-over takes place.

prophase II. A brief phase in meiosis which is similar to prophase in mitosis.

prophyll. **1.** One of the first leaves of a lateral branch. **2.** A bracteole or small scalelike appendages as in *Psilotum.*

prophyllum. The bracteole at the base of an individual flower as in *Juncus.*

proplastid. Small, self-reproducing organelles in the cytoplasm which develop into various types of plastids, e.g. chloroplasts, chromoplasts and leucoplasts.

prop roots. Adventitious roots that arise from the main trunk or stem and help to support the plant; common in many monocots, e.g. corn and *Pandanus,* and certain dicots as mangrove. *Syn.* stilt roots. (*See* fig. P-16).

prosenchyma. A type of plectenchyma in which the component hyphae lie parallel to one another.

prosorus, *pl.* **prosori.** A cell or cells which gives rise to a sorus, e.g. in the chytrid fungi.

Fig. P-16. Prop roots.

prostrate.   Trailing to lying on the ground without rooting at the nodes. *Syn.* procumbent.

protandrous, protandry.   1. Said of flowers in which the anthers mature and shed their pollen prior to the stigma of the same flower being receptive. **2.** In bryophytes, the earlier production of male gametes than female gametes in bisexual organisms.

protantherous.   Bearing leaves before flowers.

protease.   An enzyme that digests protein by hydrolysis of peptide bonds.

protein.   A complex organic compound made up of amino acids linked together by peptide bonds.

proteinaceous.   Made of or containing protein.

prothallial cell.   The vestigial vegetative cell or cells found in the microgametophytes (male gametophyte) of heterosporous vascular plants. (*See* fig. P-17).

prothallium.   *See* prothallus.

prothallus, *pl.* **prothallia.**   The more or less independent, free-living gametophyte found in ferns and other lower vascular plants. *Syn.* prothallium.

protista.   A group of organisms made up of a heterogenous assemblage of unicellular, colonial and multicellular eukaryotes.

proto-.   A prefix meaning first.

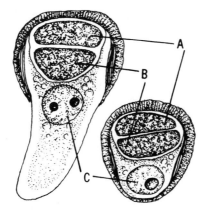

Fig. P-17. Prothallial cell: A. prothallial cell, B. generative cell, C. tube nucleus.

protocorm.   A more or less undifferentiated mass of parenchymatous tissue bearing roots or rhizoids; e. g. as in orchids and lycopods.

protoderm.   One of the three primary meristematic tissues. Protoderm differentiates into the epidermis. *Syn.* dermatogen. (*See* fig. M-2).

protogynous, protogyny.   1. Said of flowers in which the stigma becomes receptive prior to maturation of anthers and dehiscence of pollen in the same flower. **2.** In bryophytes, the earlier production of female gametes than male gametes in bisexual organisms.

protomeristem.   *See* promeristem.

proton.   A subatomic particle which has a positive charge equal in magnitude to the charge of an electron and a mass of 1.0073 mass units. It is a component of every atomic nucleus.

protonema, *pl.* **protonemata.**   The initial filamentous stage in the development of the haploid gametophyte of many mosses, liverworts, and Charophyceae (stoneworts).

protoperithecium, *pl.* **protoperithecia.**   A perithecial initial that develops into the perithecium in some sac fungi (Ascomycetes). Also called an archicarp.

protophloem.   The first differentiated elements of the primary phloem.

protoplasm.   An inclusive term for all living material of a cell.

protoplasmic streaming.   *See* cyclosis.

protoplasmodium, *pl.* **protoplasmodia.**   A microscopic plasmodium without a differentiated fan-shaped region, that produces only a single, minute sporocarp.

protoplast.   All the protoplasmic and nonprotoplasmic contents of a cell exclusive of the cell wall. (*See* fig. C-3).

protostele.   The simplest type of stele, made up of a solid column of vascular tissue with the xylem centrally located and surrounded by phloem.

protoxylem.   The first differentiated elements of the primary xylem.

provascular tissue.   *See* procambium.

proximal.   Situated near the point of reference, usually the stem; opposite of distal.

pruinose.   Having a heavy waxy bloom on the surface. More pronounced than glaucous.

prune, pruning.   The removal of stems or branches of woody plants. Objectives of pruning include controlling size and shape, and improving the quality and/or quantity of fruit and flowers.

pruniform.   Plum-shaped.

psammophyte.   A plant which grows in sand or on sandy soil.

pseud-, pseudo-.   A prefix meaning false.

pseudanthium.   An inflorescence composed of many small flowers which simulates an individual flower.

pseudoaethalium.   A cluster of fused sporangia which resemble an aethalium. Found in some slime molds (Myxomycetes).

pseudoallelism.   The phenomenon of having crossovers between the limits of what had previously been described as a single gene.

pseudobulb.   A solid, above ground, thickened or bulbiform stem, characteristic of some orchids. (*See* fig. P-18).

pseudocapillitium, *pl.* **pseudocapillitia.**   An irregular plate, tube, or threadlike body present among the spores within the fructification of some slime molds (Myxomyctes). Resembles a capillitium.

pseudocarp.   A fruit type in which an aggregation of achenes are embedded in a fleshy receptacle, e. g. strawberry.

Fig. P-18. Pseudobulb.

pseudocilium, *pl.* **pseudocilia.**   Thin, hairlike processes, e. g. in some green algae.

pseudocleistogamy.   Failure of flowers to open because of various ecological pressures, e. g. heat, moisture, etc.

pseudodrupe.   A two to four loculed nut surrounded by a fleshy involucre, e. g. walnuts.

pseudoelater.   Sterile uni- and multicellular structures found among the meiospores in a sporangium of the hornworts (Anthocerotaceae).

pseudofossils.   Various inorganic structures which resemble actual plant fossils; e. g. the mineral formations of dendrites.

pseudogamy.   The condition where pollination is necessary for seed development, even though fertilization does not occur. Found in some apomictic plants.

pseudomonopodial.   Appearing monopodial, but with the main branch not completely dominant, as in some lycopods.

pseudomycelium, *pl.* **pseudomycelia.**   A series of loosely connected cells adhering end to end forming a chain, e. g. some yeasts.

pseudoparaphysis, *pl.* **pseudoparaphyses.**   Sterile, vegetative threads found in an ascocarp. They originate at the roof of the ascocarp, grow downward and attach to the floor of the ascocarp.

pseudoparenchyma, *pl.* **pseudoparenchymata.**   **1.** A type of plectenchyma consisting of oval or isodiametric cells, which resemble parenchyma tissue when sectioned. **2.** A mass of tightly intertwined filaments which have lost their individuality and are randomly arranged.

pseudoperianth.   A layer of cells which extends from the receptacle and forms a collar-like extension around each archegonium; e. g. as in the archegoniophore of liverworts.

pseudoperithecium, *pl.* **pseudoperithecia.**   A unilocular ascostroma found in some sac fungi (Ascomycetes).

pseudoplasmodium, *pl.* **pseudoplasmodia.**   In cellular slime molds (Acrasiomycetes), a multicellular structure formed by an aggregation of amoeboid cells (myxamoebae); this structure, known as a slug, eventually produces a sporocarp.

pseudopodium.   A false foot. A stalklike extension of the gametophore of certain mosses (e.g. *Sphagnum*) which elevates the sporophyte above the gametophyte prior to the discharge of spores.

pseudoseptum, *pl.* **pseudosepta.**   A septumlike plug found in some fungi.

pseudothecia.   Ascocarps which bear a superficial resemblance to perithecia and contain bitunicate asci; e. g. as in Pleosporales.

pseudovivipary.   A method of vegetative reproduction in which vegetatively produced propagules replace some or all of the flowers in an inflorescence. (*See* fig. P-19).

psilate.   A pollen grain which appears smooth; one without visible external features; e.g. *Betula*.

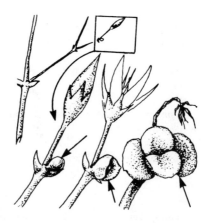

Fig. P-19. Pseudovivipary: Arrows indicate developmental sequence from left to right.

psilophytes.   Plants which live in prairies or grasslands.

psychophily.   Pollination by butterflies.

psychrometer.   An instrument used for measuring relative humidity.

pterid-.   A prefix meaning wing.

pteridophyte.   A common name referring to ferns.

ptyxis.   The rolling or folding of individual organs such as leaves and petals in the bud; also the arrangement of leaves, petals, cotyledons, etc. within a structure. *Syn.* vernation.

puberulent.   Minutely pubescent.

pubescence, pubescent.   Covered with short, soft trichomes.

puddled soil.   Clay soils which have become compacted.

pulse.   Another name for legume crops, e. g. soybeans or peas.

pulverulent.   Covered with very fine, powdery, wax granules.

pulvinate.   Cushion-shaped.

pulviniform.   Pulvinus-shaped; pad or cushion-shaped.

pulvinus.   The enlarged base of a petiole or petiolule; functions in the movement of a leaf or leaflet. (*See* fig. P-20).

puncta, *pl.* **punctae.**   Tiny pores on the valves of diatoms.

punctate.   Dotted with minute depressions or pits.

puncticulate.   Minutely punctate.

pungent.   Ending with a sharp rigid point or tip.

punnett square.   The checkerboard method used to determine the kinds of zygotes that can be produced by the fusion of gametes from the parents. Allows for the visual computation of genotypic and phenotypic ratios.

pure culture.   A culture of a single type of microoganism.

pure line.   A lineage of individuals which are homozygous at all loci, obtained by successive self-fertilizations.

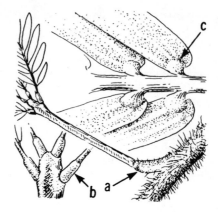

Fig. P-20. Pulvinus: a. at base of compound leaf, b. at base of primary divisions of compound leaf, c. at base of leaflet.

purine.   A nitrogenous base with a double-ring structure found in nucleic acids; e. g. adenine or guanine.

pustular, pustulate, pustulose.   Having scattered blisterlike or pimplelike elevated areas.

pycnidiospore.   A conidium (asexual spore) produced in a pycnidium.

pycnidium, *pl.* **pycnidia.**   A globose or flask-shaped structure in which conidia are formed as in some Fungi Imperfecti.

pycnosclerotium, *pl.* **pycnosclerotia.**   A more or less hard-walled structure resembling a pycnidium but containing no spores.

pycnoxylic.   Secondary xylem with few parenchymatous rays, composed mostly of conducting elements; e. g. as found in several groups of extant and extinct gymnosperms.

pyramid of energy.   A method of illustrating the energy relationships among the various feeding levels of a food chain; organisms at the bottom of the pyramid represent the greatest amount of available energy and those at the top have the least available energy. Thus, autotrophs are at the base of the pyramid; on top of these are herbivores, then primary carnivores, then secondary carnivores, etc.

pyramidal.   Pyramid-shaped.

pyrene.   The pit or seed of a drupe which is surrounded by a bony endocarp.

pyrenoid. A structure found in the chloroplasts of many algae and some bryophytes that is apparently associated with starch deposition.

pyriform. Pear-shaped.

pyrimidine. A nitrogenous base with a single-ring structure found in nucleic acids; e. g. cytosine, thymine or uracil.

pyrophyte. Plants which live in savannahs or open woodlands.

pyxidate. With a lid.

pyxis. A many seeded circumscissile capsule.

Q

quadr-, quadra-, quadri-.   A prefix meaning four.

quadrat.   A frame of any shape or size used in ecological studies for measuring various parameters of the organisms which exist in a given area; e.g. population density, cover, frequency, etc.

quadrate.   Square, or nearly square.

quadrifid.   Divided into four lobes or parts.

qualitative character.   A character in which variation is discontinuous.

quantasomes.   Tightly packed granules located on the inner surfaces of chloroplast lamellae.

quantitative inheritance.   Inheritance of genetic characters which are controlled by two or more independent genes; these genes affect quantitative characters such as growth rate, size, yield, quality, density of pigmentation and hairiness.

quantum, *pl.* **quanta.**   A unit or discrete quantity of light energy, according to the quantum theory.

quantum speciation.   The appearance of a new and very distinct species; derived from a semi-isolated peripheral population of an ancestral, cross fertile species.

quarternary leaflet.   A leaflet of the fourth degree; a leaflet of a tertiary leaflet.

quarter-sawed wood.   Wood sawed along a radial plane thereby exposing the radial surface.

quiescent center.   A region behind the root cap occupied by a group of cells that either do not divide or divide very slowly.

quilled.   Normally ligulate florets which have become tubular.

quin-, quinque-.   A prefix meaning five.

quinate.   Growing together in fives, as leaflets from the same point.

294

quincuncial.   In aestivation, the arrangement of five partially imbricated structures, e.g. sepals or petals, in which two are exterior, two interior and the fifth with one margin covering an interior structure and one margin covered by the margin of an exterior structure; e.g. as in the calyx of roses.

R

race. A genetically distinct interbreeding division of a species.

raceme. An unbranched, indeterminate inflorescence, in which the individual flowers are borne on pedicels along the main axis. (*See* fig. P-1).

racemiform. Having the form of a raceme.

racemose. Like a raceme; having flowers in racemelike inflorescences that may or may not be true racemes.

rachilla. The diminutive of rachis; in particular, the axis of the spikelet of grasses and sedges that bears the florets.

rachis. **1.** The axis of a compound leaf or fern frond upon which the leaflets are attached. **2.** The major axis of an inflorescence.

radially symmetrical. *See* actinomorphic.

radial parenchyma. *See* ray parenchyma.

radial section. A longitudinal section cut through a cylindrical body along a radius. (*See* fig. C-12).

radial symmetry. *See* actinomorphic.

radial system. The total of all rays found in secondary xylem and secondary phloem. *Syn.* horizontal system, ray system.

radiant energy. Electromagnetic radiation originating in the sun; e.g. gamma rays, ultraviolet, visible light, infrared, etc.

radiate. **1.** Spreading from, or arranged around a common center. **2.** Descriptive of a head which has ray flowers, as in many composites (Asteraceae).

radiation. The emission and propagation of energy in the form of waves.

radical. Belonging to or proceeding from the root; as in basal leaves.

radicant. Rooting; descriptive of a stem which produces adventitious roots.

radicle.   The embryonic root. (*See* fig. R-1, V-4).

radicles.   The rhizoids in mosses.

radioactive dating.   A technique used to measure the age of organic samples by determining the ratios of various radioactive isotopes, or the products of radioactive decay.

radioactive isotope.   An isotope of an element which has an unstable nucleus. It is stabilized by emitting radiation. *Syn.* radioisotope.

radioactivity.   The spontaneous disintegration of certain elements by the emission of one or more types of radiation, such as alpha rays, beta rays or gamma rays.

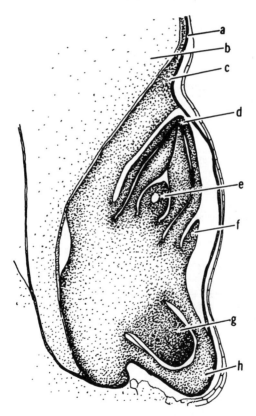

Fig. R-1. Radicle: a. aleurone layer, b. endosperm, c. scutellum, d. coleoptile, e. shoot apex, f. epiblast, g. radicle, h. coleorhiza.

radioautograph.   A photographic picture showing the location of radioactive isotopes in a tissue or other substance.

radioisotope.   *See* radioactive isotope.

rain forest.   Any temperate or tropical forest that receives over 250 centimeters of precipitation annually.

ramal, rameal.   Belonging to a branch.

rameal sheath.   A leaf sheath found at the stem joint, as in horsetails (*Equisetum*).

ramentum, *pl.* **ramenta.**   One of many thin, chaffy scales found on an epidermis; as in the scales on many fern shoots. *Syn.* ramentaceous.

ramet.   An individual member of a clone.

ramified, ramiform.   Branched; branching.

ramiform pit.   A branched pit formed by the coalescence of two or more simple pits during the deposition of the secondary wall.

ramose.   Branching or having many branches.

ramulose.   With many branches.

random.   Arrived at by chance without discrimination.

range.   1. An extensive natural pasture area. 2. The area to which a species is native; i.e. its region of natural distribution. 3. The difference between the largest and smallest values in a sample.

rank.   A vertical row.

raphe.   1. A groove in the frustule of a diatom. 2. A ridge on a seed; formed by a portion of the funiculus that is adnate to the ovule, as in an anatropous ovule.

raphide.   A needle-shaped crystal of calcium oxalate, usually occurring in bundles; e.g. as in the aroids (Araceae). (*See* fig. R-2).

rassenkreis.   A polytypic species in which distinct races replace one another in a geographical succession.

ratio-cline.   Clinal variation which occurs in a polymorphic species.

ratoon.   Regrowth of sugar cane, or the crop resulting from regrowth. *See* offset.

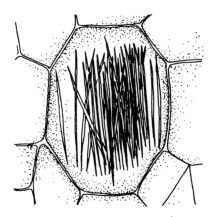

Fig. R-2. Raphides.

ray.   **1.** A tissue that extends radially into the secondary vascular tissues; formed by the ray initials in the vascular cambium. **2.** One of the primary branches of an umbel. **3.** A ray flower.

ray flower.   A ligulate, zygomorphic flower found in the inflorescence (head) of many composites (Asteraceae). Contrast disk flower. (*See* fig. C-2).

ray initial.   A meristematic cell in the vascular cambium that gives rise to ray cells in the secondary vascular tissue.

ray parenchyma.   Parenchyma cells of a ray, in contrast with axial parenchyma.

ray system.   *See* radial system.

ray tracheid.   A tracheid cell occurring in a ray in the secondary xylem of certain conifers.

reaction wood.   Wood with distinctive anatomical characteristics formed in leaning or crooked stems. Formed on the lower side of a branch in conifers and on the upper side of a branch in dicotyledons. *See* compression wood and tension wood.

recapitulation theory.   The theory that as an organism develops, it passes through stages that resemble the adult forms of its successive ancestors. Often stated as "ontogeny recapitulates phylogeny".

receptacle.   **1.** In angiosperms the portion of the axis of a flower stalk on which the flower is borne. (*See* fig. C-2). **2.** In certain liverworts, a stalked structure that bears the antheridia and archegonia.

recessive, recessive gene.   In genetics, a gene whose phenotypic expression is masked by a dominant allele when in a heterozygous condition.

reciprocal crosses.   Crosses in which the sources of male and female gametes are reversed.

reclinate, reclining, reclined.   Turned or bent downward away from perpendicular.

recombinant.   The individual or cell resulting from recombination.

recombination.   The formation of new combinations of genes not found in the parents. Results from independent assortment and/or the rearrangement of linked genes due to crossing-over.

recon.   The smallest unit of DNA capable of recombination.

recurved.   Bent or curved downward or backward.

red algae.   Algae of the division Rhodophyta.

redifferentiation.   A reversal in differentiation in a cell or tissue and subsequent differentiation into another type of cell or tissue.

red tide.   A population explosion of certain marine dinoflagellates, e.g. *Gonyaulax, Glenodinium* and *Gymnodinium;* these occur off the coasts of California and Florida, resulting in a reddish hue of daytime water and luminescence at night; they are poisonous to many marine animals.

reduced.   Secondarily simplified during evolution.

reductant.   The molecule that is oxidized by the loss of an electron in an oxidation-reduction reaction.

reduction.   A gain of an electron by an atom or molecule which occurs simultaneously with an oxidation reaction, i.e. a loss of an electron by an atom or molecule.

reduction division.   *See* meiosis.

redundant DNA.   *See* polygenes.

reflectivity.   The ratio of the radiant energy reflected from a surface to the total incident radiant energy.

reflexed.   Abruptly bent or recurved downward or backward.

reforestation. Replanting a forest.

refractile. Capable of reflecting.

refulgens. Shining brightly.

regeneration. The regrowth of new tissues, organs, or fragments of them, by an organism.

region of cell division. The region behind the root cap in which cell division takes place. (*See* fig. M-2).

region of elongation. The area behind the region of cell division where cell elongation takes place. (*See* fig. M-2).

region of maturation. The region where cellular differentiation, e.g. root hair development, takes place. *Syn.* region of differentiation. (*See* fig. M-2).

registered seed. The progeny of foundation seed normally grown to produce certified seed.

regression coefficient. A numerical measure of the rate of change of the dependent variable with respect to the independent variable.

regular. *See* actinomorphic.

regulator gene. A gene that controls the activity of the structural genes in an operon, by the production of a protein repressor. (*See* fig. O-2).

releve. A sample stand of vegetation.

relict. A plant with a restricted geographical distribution; a remnant from an earlier geological period.

remote. Scattered; widely separated.

renaturation. The return to its native, three-dimensional configuration from a denatured state; as with a denatured nucleic acid.

reniform. Kidney-shaped; having broadly rounded margins and a shallow sinus. (*See* fig. R-3).

repand. Having a shallowly undulating or sinuate margin.

repent. Creeping along the ground and rooting at the nodes.

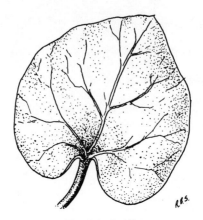

Fig. R-3. Reniform.

replicate.   1. To produce a facsimile or duplicate from an original or template. **2.** Folded abaxially along the middle, as along the midrib of a leaf.

replum.   A persistent partition or septumlike placenta which bears ovules on the margins and separates the valves in the fruiting carpels in mustards (Brassicaceae). (*See* fig. S-7).

repressor.   *See* regulator gene. (*See* fig. O-2).

reproduction.   The formation of a new individual by sexual or asexual means.

reproductive isolation.   The prevention of interbreeding of two groups of organisms by a variety of mechanisms.

repulsion.   Denotes a genetic linkage in which each homolog contains a mutant and a wild-type gene. *Opp.* coupling.

reservoir.   An enlarged portion of the gullet of some motile cryptomonad algae (Cryptophyceae).

residual meristem.   A tissue that is the least differentiated portion of the apical meristem. It gives rise to the procambium and the interfascicular ground tissue.

resin canal.   *See* resin duct.

resin duct.   A tubelike, intercellular cavity containing resin and lined by resin-secreting cells; as in the conifers. *Syn.* resin canal. *See* epithelial cell. (*See* fig. H-6).

resinous, resiniferous.   Producing or containing resin.

resistance.   The ability of a host to suppress or retard the activity of a pathogen.

respiration.   An intracellular series of enzyme-mediated chemical reactions in which food is oxidized and energy is released. The complete breakdown of the organic compounds also produces carbon dioxide and water.

resting nucleus, resting cell.   A nucleus or cell which is not undergoing division, but which is usually very active metabolically.

resting spore.   A thick-walled dormant spore.

resupinate.   Appearing to be upside down; twisted 180° from the normal orientation.

reticulate.   Forming a network pattern; e.g. as in reticulate cell wall thickenings, reticulate sieve plate, reticulate perforation plate, etc.

reticulate evolution.   Evolution in which phyletic lines form a network.

reticulate venation.   *See* net venation.

reticulum.   A network, or netlike structure.

retinaculum.   **1.** The persistent, hooklike funiculus, as in the fruits of the Acanthaceae. **2.** The structure to which the pollinium is attached, as in the milkweeds (Asclepiadaceae) or the orchids (Orchidaceae).

retort cell.   A flask-shaped cell having an apical pore; as in some mosses.

retrorse.   Bent or directed downward or backward. *Opp.* antrorse.

retting.   The process used to free fiber bundles from other tissues by using microorganisms to dissolve thin-walled cells which surround the fibers; used on stems and leaves, especially with many monocots.

retuse.   Having a shallow notch in a rounded or obtuse apex.

reverse mutation.   A change of a mutant gene back to its original state.

revolute.   With the margins rolled downward, or toward the lower side. *Opp.* involute.

rexigenous.   Refers to an intercellular space which has originated by rupture of cells. Compare lysigenous and schizogenous.

R$_f$.   In paper chromatography, a ratio of the distance a solute travels divided by the distance traveled by the solvent. The solvent must be specified for each R$_f$ value.

rhipidium.   A more or less fan-shaped cyme.

rhizanthous.   Flowering or appearing to flower from the root.

rhizine.   A bundle of hyphae that attaches a lichen thallus to a substrate; also absorptive in function.

rhizocarp.   *See* sporocarp.

rhizocarpic, rhizocarpous.   Refers to plants that have perennial roots but with stems that die back annually.

rhizodermis.   The outer surface layer of a root, not homologous with a shoot epidermis.

rhizoid.   A rootlike structure in function and appearance, but lacking vascular tissue; **1.** In fungi and algae, rootlike structures which absorb water, food and nutrients. **2.** In liverworts, mosses and some vascular plants, roothair-like structures usually found only on the gametophyte.

rhizoidophores.   A single bulbous cell which is laterally inserted on an endophytic hyphal filament, and attached to it by a single stalk cell.

rhizomatous.   Producing or bearing rhizomes; like a rhizome.

rhizome.   An underground, more or less horizontal stem. (*See* figs. H-5, R-4).

rhizomorph.   A thick strand of somatic hyphae composed of many individual hyphae; the whole mass functions as an organized unit.

rhizomycelium, *pl.* **rhizomycelia.**   An extensive rhizoidlike system; superficially resembles mycelium.

rhizophores.   Proplike appendages that grow downward from the ascending stems of some club mosses, e.g. *Selaginella;* they branch and become rootlike upon entering the ground; sometimes bear leaves and cones.

rhizoplast.   Chromatic strands that connect the blepharoplast to the nucleus.

rhizopodial.   A growth form which is somewhat amoeboid.

rhizosphere.   The region in the soil surrounding the roots, especially the absorptive region; this region receives metabolites from the root and supports a unique microflora.

rhizotaxis.   The position or arrangement of lateral roots.

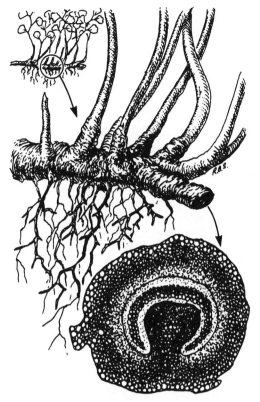

Fig. R-4. Rhizome.

rhombate, rhomboidal.   Having a rhombic shape.

rhombic.   More or less diamond-shaped; having straight margins and being widest in the middle.

rhytidome.   The outer bark; the periderm, and associated cortical and phloem tissues.

rib.   An elongated ridge or protrusion; as in the main veins of a leaf.

rib meristem.   A meristematic tissue giving rise to parallel or vertical files of cells. *Syn.* file meristem.

ribonucleic acid (RNA).   A nucleic acid formed on a DNA template, and involved in protein synthesis; RNA occurs in three forms. *See* messenger RNA, ribosomal RNA and transfer RNA.

ribose. A five-carbon sugar; a component of RNA.

ribosomal RNA (rRNA). The type of RNA found in the ribosome.

ribosome. A cellular granule (or particle) composed of protein and RNA, and functions in protein synthesis; they occur in the nucleus, mitochondria, plastids and are found free in the cytoplasm, or attached to the endoplasmic reticulum.

ring. *See* inner veil. *Syn.* annulus. (*See* fig. B-2).

ring chromosome. **1.** A ring-shaped chromosomal association seen during diakinesis. **2.** An aberrant chromosome without ends.

ringed bark. A split or cracked bark which has circular fissures.

ringent. Gaping open; as the mouth of an open-throated bilabiate corolla.

ringing. *See* girdling.

ring porous wood. Wood in which the vessels or pores are larger in the early wood than in the late wood; forms a well-defined ring as seen in cross sections of the wood. (*See* fig. D-3).

riparian. Growing by rivers or streams.

RNA. *See* ribonucleic acid.

robust. Large; healthy.

rogue. A plant that does not come true from seed.

roguing. The process of removing undesirable individuals from a group of plants, in order to maintain purity; e.g. the removal of weeds from a seed producing plot.

rolled. Sides enrolled, usually loosely.

root. The descending axis of a plant, normally below ground; functions include anchorage, absorption and conduction of water and minerals, and sometimes food storage; roots lack nodes and internodes.

root cap. A parenchymatous, dome-shaped mass of cells at the tip of a root, formed by and protecting the apical meristem. (*See* fig. M-2).

root hair. A slender, lateral outgrowth of an epidermal cell of a root, which serves to extend the absorbing surfaces of the root; most root hairs are short-lived and are usually confined to the region (or zone) of maturation. (*See* fig. M-2).

root hair zone.   *See* region of maturation.

rooting medium.   *See* medium.

rootlet.   A small root.

root nodule.   A swelling on the roots of legumes that are produced by symbiotic nitrogen-fixing bacteria. (*See* fig. R-5).

root pressure.   The pressure developed in roots as a result of osmosis; causes guttation, and in cut stems, causes exudation.

root rot.   Referring to various diseases of roots usually caused by fungi.

rootstock.   A rhizome.

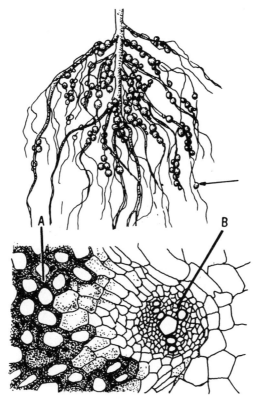

Fig. R-5. Root nodule: A. bacteria filled nodule cells in cross section, B. vascular bundle of host root in cross section.

root tip.   The apex of a young root.

roridulate.   Covered with waxy platelets, thereby appearing dewy.

rosette.   A crowded, circular cluster of leaves or other organs; often in reference to a growth habit in which leaves radiate from a crown, close to the ground. (*See* fig. R-6).

rosette cells, rosette tier.   The basal tier of cells of the proembryo. (*See* fig. S-4).

rostellate.   Minutely rostrate.

rostellum.   **1.** A small beak. **2.** A sterile modified stigmatic lobe in some orchids.

rostrate, rostrum.   Having a beak.

rosulate.   Having the form of or being in a rosette.

rotate.   Spreading, or saucer-shaped; descriptive of a gamopetalous corolla which has a short tube and a widely spreading circular limb, as in many members of the nightshade family (Solanaceae).

rotation.   The planting of different crops one after another for maximum utility.

rotund.   Round or nearly circular in outline.

rough ER.   Endoplasmic reticulum with ribosomes attached to its surface. Contrast smooth ER.

rRNA.   *See* ribosomal RNA.

Fig. R-6. Rosette.

r selection. Selection favoring a high population growth rate and high productivity. Characteristic of colonizing species. Compare K selection.

rubellous, rubescent. Reddish or turning red.

rubiginose. Rust colored.

ruderal. Weedy; a plant growing in disturbed habitats, or in waste places.

rudimentary. Imperfectly or incompletely developed; vestigial and usually nonfunctional.

ruffled. Having a very strongly wavy margin.

rufous. Reddish-brown.

ruga, *pl.* **rugae.** A wrinkle or fold.

rugose. Wrinkled; covered with coarse reticulate lines.

rugulate. Wrinkled or wormy.

rugulose. Finely wrinkled.

ruminate. Having a chewed appearance; coarsely wrinkled.

runcinate. Sharply or coarsely pinnatifid or cleft, with the jagged lobes pointing downward toward base, as in a dandelion leaf.

runner. A specialized stem (stolon) which develops from the axil of a leaf at the crown of a plant, has long internodes and grows horizontally along the ground; it forms adventitous roots and shoots at some of the nodes.

rupturing. Bursting irregularly.

rush. Any member of the rush family (Juncaceae).

russet. A brownish roughened area on fruit skins, from the abnormal production of cork tissue caused by disease, insect, or spray injury.

rust. A fungus in the Uredinales which causes rust diseases.

S

S₁, S₂, S₃, etc. Symbols used to represent a series of selfed (self-fertilized) generations; e.g. S_1 denotes the generation obtained when the parent is selfed.

sac. A bag-shaped structure, pouch or indentation.

saccate. Having the shape of a sac or pouch.

saccus, *pl.* **sacci.** The winglike extensions of the exine on certain pollen grains; these give buoyancy to wind disseminated pollen, as in conifers.

sac fungi. A common name for fungi in the class Ascomycetes.

sagittal section. A section cut perpendicular to the surface; a cut parallel to the long axis of a dorsoventral structure.

sagittate. Shaped like an arrowhead; triangular-ovate with two straight or slightly concave basal lobes.

salient. Pointing outward, as teeth from a leaf margin.

salinity. The relative concentration of salts in water or soil; especially in reference to sodium chloride.

saline soil. A soil containing a large amount of soluble salts.

salt marsh. Any marsh situated in saline water; may be inland or along the seacoast.

salverform. Shaped like a trumpet. Said of a sympetalous corolla which has a slender tube and an abruptly expanding flat limb, e.g. phlox. *Syn.* hypocrateriform and salver-shaped.

salver-shaped. *See* salverform.

samara. A one- or two-seeded, simple, dry, indehiscent fruit. The fruit pericarp bears winglike outgrowths that aid in wind dissemination, e.g. ash (*Fraxinus*).

samaracetum. A group of samaras; e.g. as in the tulip poplar tree (*Liriodendron*).

sample. A subset of measurements taken from selected individuals assumed to be representative of a population.

310

sampling error.   Deviation in a sample value due to the limited size of the samples.

sanguineous.   Blood-red; crimson.

sap.   The fluid contents of xylem or phloem. *See* cell sap.

sapid.   Having a pleasant taste.

saponaceous.   Slippery or soapy to the touch.

saprobe.   A saprophyte; an organism that gets its food from nonliving organic matter.

sapromyophily.   Pollination by carrion or dung flies. Also spelled sapromyiophily.

saprophyte.   A plant that obtains its food from nonliving organic matter.

saprophytic.   Obtaining food from nonliving organic matter.

sapwood.   The outer portion of the wood (xylem) of a stem or trunk in which active conduction of water occurs. It is normally lighter in color than the inner portions (heartwood) of the wood in a stem or trunk. (*See* fig. C-12).

sarcocauly.   Having fleshy stems.

sarcous.   Having a fleshy seed coat.

sarment.   A long, slender, prostrate runner or stolon.

sarmentose.   Bearing long, slender, prostrate runners or stolons.

satellite.   A small, terminal segment of a chromosome which is separated from the rest of the chromosome by a conspicuous secondary constriction. (*See* fig. S-1).

savanna.   A grassland with scattered trees or scattered clumps of trees; the vegetation is able to withstand hot and dry seasons of considerable length.

Fig. S-1. Satellite.

saxicolous.   Living or growing among rocks.

scab.   A crustlike disease lesion; a disease in which scabs are prominent symptoms.

scaberulent.   Nearly scabrous.

scaberulous, scabrellate.   Diminutive of scabrous.

scabrate pollen.   Pollen with ridges or processes (scabrae) of different shapes, smaller than one micrometer, e.g. *Quercus*.

scabrid.   Roughened.

scabridulous.   Slightly rough; minutely scabrous.

scabrous.   Having a surface that is rough to the touch, because of the presence of short stiff hairs.

scalariform.   Having ladderlike markings.

scalariform cell wall thickening.   Secondary wall material deposited in a ladderlike pattern in vessel elements and tracheids. *Syn.* scalariform thickening.

scalariform conjugation.   Refers to the ladderlike appearance created by the formation of conjugation tubes between two laterally adjacent filaments, as in some algae, e.g. *Spirogyra*.

scalariform perforation plate.   A type of multiperforate plate in a xylem vessel member formed by elongated perforations arranged parallel to one another in a ladderlike pattern.

scalariform pitting.   Elongated pits which are arranged parallel to one another forming a ladderlike pattern. Found on the tracheary elements of xylem. (*See* fig. P-9).

scalariform-reticulate cell wall thickening.   A secondary wall thickening on some tracheary elements which is intermediate between scalariform and reticulate.

scalariform sieve plate.   A compound sieve plate in phloem formed by elongated sieve areas arranged parallel to one another in a ladderlike pattern.

scalariform thickening.   *See* scalariform cell wall thickening.

scale.   **1.** Any thin, usually small and dry, scarious to coriaceous bract. **2.** A usually disk-shaped trichome attached by a stalk. *Syn.* peltate hair.

scale leaf.   Small leaves resembling scales.

scandent.   Climbing.

scape.   A leafless flowering stalk arising from the ground level of an acaulescent plant. Stalk may bear scales or bracts but has no foliage leaves. (*See* fig. S-2).

scaphoid.   Shaped like a boat.

scapiform.   In the form of a scape.

scapose.   Bearing one or more flowers on a scape; in the form of a scape.

scar.   **1.** A mark left by the natural separation of one organ from another; e.g. as on a stem after abcission of a leaf, bud, flower or fruit, or on a seed after its detachment from a fruit. (*See* fig. B-8). **2.** A mark left after regrowth of damaged tissue following an injury.

scarification.   The act of scarifying seeds; any process which renders the seed coat permeable to water and gas exchange, in order to hasten germination; e.g. breaking or scratching of the seed coat or treatment with hot water or acids.

scarious.   A thin, nongreen, dry, membranaceous structure.

scatter diagram.   A graphic means of showing the relationship between two characteristics. Correlation may be positive, negative or absent.

schizo-.   A prefix meaning split.

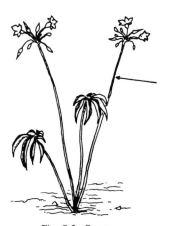

Fig. S-2. Scape.

schizocarp. A dry, dehiscent fruit that splits into two or more separate one-seeded carpels (mericarps) at maturity. Derived from an inferior ovary. Examples include schizocarpic achenes, berries, follicles, nutlets, samaras, etc.

schizogenous. The manner of formation of intercellular spaces by the separation of cells along their middle lamella. Compare lysigenous and rexigenous.

schizo-lysigenous. The manner of formation of intercellular spaces by both cell separation along their middle lamellae and by cell disintegration.

schizopetalous. Having divided or split petals.

scientific method. A method of problem solving that consists of the following steps: (1) awareness of a problem; (2) delineation of one or more hypotheses formulated to explain the problem; (3) testing these hypotheses by observation and experimentation; and (4) accepting or rejecting the hypotheses on basis of evidence.

scion. A portion of a stem that is transferred to a new root stock in grafting. *Syn.* cion.

sciophyll. A shade leaf.

sciophyte. A plant growing in the shade.

scissile. Easily split or separated.

scler-, sclero-. A prefix meaning hard.

sclereid. A sclerenchyma cell having a thick, lignified secondary cell wall with many pits; they may or may not be living at maturity, are variable in form, but are usually not elongated. Compare fiber. (*See* fig. S-3).

sclerenchyma. A tissue composed of sclerenchyma cells, including fibers, fiber-sclereids and sclereids; functions include support and sometimes protection.

sclerenchyma cell. A supporting cell, variable in form and size, having a more or less thick, often lignified secondary cell walls. Includes fibers, fiber-tracheids and sclereids.

sclerenchyma fiber. *See* fiber.

sclerenchymatous. With reference to sclerenchyma tissues or cells.

sclerification. The process of changing into sclerenchyma through the formation of secondary walls.

sclerophyll. A firm, leathery or tough leaf consisting of a large amount of sclerenchyma; such leaves are characteristic of many perennial plants adapted to dry habitats.

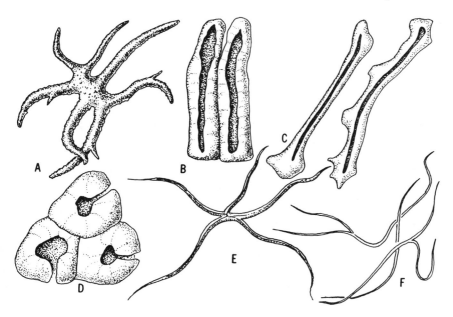

Fig. S-3. Sclereids: A. astrosclereid, B. macrosclereid, C. osteosclereid, D. brachysclereid, E. trichosclereid, F. filiform sclereid.

sclerotic.   Having sclereids.

sclerotic parenchyma cell.   A parenchyma cell that becomes a sclereid through deposition of a thick secondary wall.

sclerotic ray cell.   A ray cell with thick, often lignified secondary walls.

sclerotium, *pl.* **sclerotia.**   A firm, dormant stage found in certain fungi; e.g. the hard, resting stage in noncellular slime molds (Myxomycetes), or a dormant stage made up of a firm mass of hyphae in some sac and club fungi.

sclerous.   Hard.

scobiform.   Looking like sawdust.

scobina.   A zigzag, rasplike rachilla of the spikelet of certain grasses.

scobinate.   Having a surface that feels rough.

scolecospore.   An elongated, needle- or wormlike spore found in some Fungi Imperfecti (Deuteromycetes).

scorch. Burning of tissue from infection or weather conditions.

scoring. A horticultural technique for inducing early fruiting of certain tree species, by cutting a number of lines in the bark in a cross-hatched pattern.

scorpioid. Circinately coiled at the tip.

scorpioid cyme. A determinate, circinately coiled inflorescence. Flowers are two-ranked and borne alternately on opposite sides of the inflorescence axis.

scrobiculate. Pitted; marked by minute or shallow depressions.

scrotiform. Pouch-shaped.

scrub. Vegetation characterized by stunted trees or shrubs.

scurfy. Covered with minute, branlike scales; with scaly incrustations.

scutate. Shaped like a small shield.

scutellate. Shaped like a small platter.

scutelliform. Platter-shaped.

scutellum. **1.** The single cotyledon of a grass (Poaceae) embryo, specialized for the absorption of food from the endosperm. (*See* fig. R-1). **2.** A plate or shieldlike cover, as in certain fungi (Microthyriales).

scutum. The dilated apex of the style in milkweeds (Asclepiadaceae).

secondary body. *See* secondary plant body.

secondary cell wall. A layer formed from a variety of substances which are deposited on the inside of, and over the primary cell wall. Consists primarily of cellulose but may or may not have other materials associated with it.

secondary endosperm nucleus. The nucleus formed as a result of the fusion of the two polar nuclei in the embryo sac of angiosperms.

secondary growth. In plants, an increase in thickness (girth) resulting from formation of secondary tissues by lateral or secondary meristems, especially the vascular cambium. Compare primary growth.

secondary leaf. A leaf produced above the cotyledon(s).

secondary meristem.   A meristem, such as the vascular or cork cambium that produces secondary plant growth.

secondary nucleus.   In certain angiosperms, the nucleus formed by the fusion of the polar nuclei prior to being fertilized by a sperm.

secondary phloem.   Phloem formed by the vascular cambium during secondary growth in a vascular plant.

secondary phloem fiber.   A fiber located in the axial system of secondary phloem.

secondary plant body.   That part of the plant body which is added to the primary body by meristematic activity. Consists of secondary vascular tissues and periderm.

secondary root.   A lateral root derived from the pericycle of a primary root. *See* branch root.

secondary succession.   An ecological succession which begins after some disturbance has destroyed the existing plant community.

secondary suspensor.   In certain gymnosperms, cells formed between the primary suspensor and the basal tier of proembryonal cells. *Syn.* embryonal tube. (*See* fig. S-4).

secondary tissue.   Tissues produced by the secondary meristems; e.g. vascular cambium and cork cambium. Compare primary tissue.

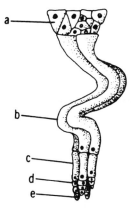

Fig. S-4. Secondary suspensor: a. rosette cells, b. primary suspensor, c. secondary suspensor, d. secondary series of embryonal tubes, e. apical cell.

secondary vascular tissue.   Xylem and phloem found by the vascular cambium during secondary growth. Differentiated into axial and ray systems.

secondary wall.   *See* secondary cell wall.

secondary xylem.   Xylem tissue formed by the vascular cambium during secondary growth in a vascular plant.

secretion.   The elimination of substances out of a cell, which may still take part in metabolic processes; in plants, a clear distinction cannot be made between secretion and excretion.

secretion vesicles.   *See* Golgi body.

secretory cavity.   A more or less spherical shaped cavity containing secreted substances derived from the cells that broke down in the formation of the cavity; such cavities are usually lysigenous in origin and occur in various and tissues. Compare secretory duct.

secretory cell.   A living plant cell specialized to secrete one or more organic substances; includes internal and external cells and covers a variety of secretory structures, e.g. simple cells, trichomes, glands, etc.

secretory duct.   Elongated spaces (ducts) which contain secretions derived from the epithelial cells lining the duct; such ducts may be lysigenous or schizogenous in origin and occur in various organs and tissues. Compare secretory cavity. *Syn.* secretory canal.

secretory hair.   *See* glandular hair.

seculate.   *See* falcate.

secund.   Turned to one side by twisting or torsion; often used in reference to an arrangement of flowers on one side of an inflorescence axis.

sedimentation coefficient.   The rate at which a given solute molecule, which is suspended in a less dense solvent, sediments as a result of centrifugal force. *Syn.* Svedberg unit or S value.

sedoheptulose.   A seven-carbon sugar.

seed.   A ripened ovule which develops following fertilization in seed plants. (*See* fig. A-7, D-6, V-4).

seed coat.   The outer protective layer of a seed which develops from the integuments of the ovule. *Syn.* testa.

seed leaf.   A cotyledon.

seedling.   A young plant developing from a germinating seed.

seed plant.   A plant that produces seeds, i.e. a spermatophyte; includes all gymnosperms and angiosperms.

seed stalk.   *See* funiculus.

seepage.   The percolation of water through the soil.

segment.   A portion or division of a plant organ, which is divided but is not compound.

segmental allopolyploid.   An allopolyploid in which certain chromosomal segments of the parental species are homologous and others are not; that is, part of the chromosomal complement is essentially autopolyploid in origin and the rest is allopolyploid.

segregation.   The separation of homologous chromosomes (and with them the genes) during meiosis. *See* Mendel's principle of segregation.

selection.   *See* artificial selection and natural selection.

selection coefficient.   A measure of the disadvantage of a given genotype in a population.

selection pressure.   The effect of natural selection in changing the genetic composition of a population over a number of generations.

self, selfed.   To be self-pollinated and/or self-fertilized.

self-compatible, self-compatibility.   An organism that is capable of reproducing sexually by itself; the ability to produce fruits with normal seeds following self-pollination.

self-fertility.   Capable of self-fertilization.

self-fertilization.   A condition in which the male (sperm) and female (egg) gametes from the same plant fuse and produce a viable zygote.

self-incompatible, self-incompatibility.   An organism that is self-sterile; i.e. it is not capable of producing seed without being cross-pollinated.

self-pollination.   The transfer of pollen from an anther to a stigma of the same flower or of another flower on the same plant, or from one flower to another within a clone.

self-sterility.   A self-incompatible organism.

semaphylls.   Floral parts (petals, sepals or tepals) with a primary function of pollinator attraction.

semataxis.   The arrangement of semaphylls on a plant and within a flower.

semen, *pl.* **semines.**   A seed.

semi-.   A prefix meaning half.

semicarpous.   Having ovaries of adjacent carpels partly fused but with stigmas and styles separate.

semicell.   One-half of a desmid. (*See* fig. S-5).

semiconservative replication.   Refers to the method by which DNA is duplicated; in the process, DNA replication occurs by longitudinal division of the molecule with each resulting half being conserved, and acting as the template for the formation of a new DNA strand.

semidominance.   A condition in which an intermediate phenotype is produced in individuals heterozygous for a particular gene.

semiglobose.   Half-globose; hemispherical.

semilethal mutation.   A mutant which causes the death of greater than 50 percent of the progeny, although not all individuals with the mutant genotype die.

semilunate.   Shaped like a half-moon, crescent-shaped.

seminal roots.   In the embryo of grasses, adventitious root primordia that develop on the hypocotyle.

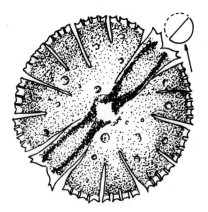

Fig. S-5. Semicell.

semipermeable membrane.   A membrane that is permeable to water but not to larger solutes, e.g. proteins. *See* differentially permeable membrane.

semisterility.   Descriptive of the results of a cross in which one-half or more of all zygotes formed are inviable.

senescence.   The aging process.

senescent.   Aging.

sensing element.   A body found closely associated with a flagella and an eyespot in euglenids. Thought to be sensitive to light and play a role in initiation of flagellar movement.

sepal.   One of the outermost, sterile appendages of a flower which normally encloses the other floral parts in the bud; one of the separate parts of the calyx. *See* calyx.

sepaloid.   Sepallike in shape, texture and/or color.

separation layer.   *See* absciss layer.

septal nectary.   A nectary which occurs in the partitions of the ovary where the carpel walls are incompletely fused; frequently found in monocots.

septate.   Being divided or partitioned by cross walls (septa) into locules or cells.

septate fiber.   A fiber which forms thin septa (transverse walls) after the secondary wall is laid down.

septicidal.   Dehiscence of a capsule through the septa between the locules. Compare loculicidal.

septum, *pl.* **septa.**   A dividing cross wall or partition.

sere.   A group of plant communities that successively occupy the same area from the pioneer stage to a mesic climax.

seriate.   In series or rows.

sericeous.   Having long, silky, slender hairs which are usually appressed. *Syn.* sericate.

serology.   The study of immunological phenomena; in plants, used to analyze the relationships among proteins in different taxa.

serophyte.   A xerophyte.

serotinal, serotinous.   Being produced, opening or appearing late in the season; occurring late in summer.

serpentine.   A soil composed primarily of hydrated magnesium silicate, and with a low concentration of calcium. It may be green, yellow or brown in color, often mottled with red.

serrate.   Having a saw-toothed margin with sharp teeth pointing forward or toward the apex.

serrulate.   Minutely or finely serrate.

sessile.   Without a stalk; sitting directly on its base.

seta, *pl.* **setae.**   **1.** A bristlelike hair. **2.** In some bryophytes, the stalk of the sporophyte that supports the capsule; e.g. in some mosses and liverworts.

setaceous.   Having bristlelike hairs; bristly.

setiferous.   Having bristles.

setiform.   Bristle-shaped.

setose.   Covered with bristles; bristly.

setulose.   Bearing minute bristles.

sex cells.   *See* gamete.

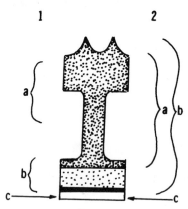

Fig. S-6. Sexine: 1. Erdtman system: a. sexine, b. nexine, c. intine. 2. Faegri system: a. ektexine, b. exine, c. intine.

sex chromosome.   Homologous chromosomes that are morphologically distinct from autosomes, and are involved in sex-determination.

sex genes.   In bisexual plants, genes which selectively depress the development of male or female sex organs; possibly through regulation of auxin level.

sexine.   The outer layer of the exine of a pollen grain; the sculptured part of exine. *Syn.* ektexine. (*See* fig. S-6).

sex-limited character.   A phenotype that is expressed in only one sex.

sex-linked.   Referring to a genetic characteristic which is determined by a gene or genes, which are located on the sex chromosomes.

sex organs.   An organ that produces gametes.

sexual.   Pertaining to sex.

sexual fertilization.   The union or fusion of two haploid gametes to produce a single zygote.

sexual propagation.   The reproduction of plants by seed.

sexual reproduction.   Reproduction involving meiosis and syngamy; the union of gametes from two parental cells.

sexual system.   An artificial system of classification devised by Linnaeus which was based upon the number and distribution of the sexual organs of a flower.

shade house.   *See* lathhouse.

shade leaves.   Leaves that are adapted to low light levels; are usually thinner and larger than leaves found on the same plant exposed to the sun.

shape.   A specific two-dimensional form or a specific three-dimensional figure.

sheath.   **1.** Any more or less tubular structure surrounding an organ or plant part. **2.** A tissue layer that surrounds another tissue or tissues; e.g. bundle sheath.

sheathing.   Enclosing, covering or surrounding.

sheathing base.   A leaf base that surrounds a stem.

shell zone.   A shelllike zone of parallel curving layers of cells found in axillary bud primorida.

shield. Any structure that is shield shaped; the thickened, rhomboid shaped extremity of a conifer cone scale.

shoot. **1.** A collective term which includes all aerial portions of a vascular plant. **2.** A young growing branch or twig with its leaves.

shoot apex. The terminal part of a shoot directly above the uppermost leaf primordia, in which the apical meristem is situated. *See* apical meristem. (*See* fig. R-1).

shoot-root ratio. The ratio of the weight of the shoot divided by the weight of the root.

short-day plant. Plants which flower only after receiving illumination that is shorter than a "critical photoperiod". Varies in length (approximately 8–10 hours), depending on the species. Compare long-day plant.

short shoot. *See* spur.

shrub. A perennial woody plant usually with several main stems arising from or near the ground; a bush.

sib. An abbreviated form of sibling.

sibling. Offspring of the same parents.

sibling species. Species that are nearly identical morphologically, but are reproductively isolated from one another.

sib mating. The mating between siblings.

sieve area. An end or side wall area of a sieve element which has pores, through which the protoplasts of vertically or laterally adjoining sieve elements are interconnected; an end wall usually has larger pores and is called a sieve plate.

sieve cell. A long, slender sieve element with undifferentiated sieve areas and tapering end walls which lack sieve plates. Typical in phloem of gymnosperms and lower vascular plants.

sieve element. A cell of the phloem concerned mainly with the longitudinal transport of food materials. Classified into sieve cells and sieve tube members.

sieve plate. A sieve area in the end wall of a sieve element, typical of angiosperm. (*See* fig. C-10).

sieve tube. A vertical series of sieve tube members with sieve plates interconnected end-on-end.

sieve tube element. *See* sieve tube member.

sieve tube member. One of the cells of a sieve tube, often associated with one or more companion cells; typical in phloem of angiosperms. *Syn.* sieve tube element.

sigmoid. S-shaped, doubly curved.

sigmoid curve. An S-shaped curve.

significance, test of. A statistical test designed to distinguish between differences due to sampling error and differences due to discrepancy between observation and hypothesis. Significance based on probability values.

silage. Forage that is chemically changed and preserved by fermentation.

silica cell. Cell impregnated with silica; e.g. epidermal cells of certain grasses.

siliceous, silicious. Composed of or impregnated with silica (silicon dioxide).

silicle. A dry, dehiscent fruit, that is as wide or wider than long; the characteristic fruit of many species in the mustard family (Brassicaceae). *Syn.* silicule. (*See* fig. S-7).

silicolemma. In certain diatoms, a flattened, surface sac in which silicon is deposited; e.g. in the plates of the "armored" dinoflagellates.

silique. A dry, dehiscent fruit, that is two-celled. The valves split from bottom up and leave the placentae with the false partition (replum) between them. Fruit narrow and longer than wide; characteristic of many species in the mustard family (Brassicaceae). (*See* fig. S-7).

Fig. S-7. Silicle (left) and silique (right): a. replum, b. funiculus, connecting placenta and seed.

silky. *See* sericeous.

Silurian. A period in the Paleozoic era which lasted from 400 to 440 million years before present.

silvery. With a whitish, metallic, more or less shining luster.

simple. Composed of not more than one anatomically or morphologically identical unit; not compound.

simple fruit. A fruit derived from a single pistil which may consist of one carpel or several united carpels.

simple leaf. An undivided leaf. Compare compound leaf.

simple perforation plate. A perforation plate which has a single perforation; found in vessel members of xylem. Contrast multiperforate perforation plate.

simple pistil. A pistil composed of only one carpel.

simple pit. A pit which is not encircled by an overarching border of secondary wall. Compare bordered pit.

simple pit-pair. An intercellular pairing of two simple pits.

simple polyembryony. The condition in which each of the fertilized eggs of one gametophyte produce separate but undivided embryos; a common occurrence in conifers, in which each proembryo produces a single embryo and the four apical cells function as a unit without cleaving.

simple sieve plate. A sieve plate having only one sieve area.

simple sorus. A sorus in which all the sporangia mature at the same time.

simple strobilus. A strobilus made up of an axis bearing lateral appendages (sporophylls) that bear sporangia; e.g. the pollen cones of conifers.

single cross. A cross between two inbred lines.

sinistrorse. A counterclockwise (i.e. turned to the left) helical growth pattern characteristic of some twining stems.

sink. A region of a plant which utilizes sugars that were translocated there by the phloem.

sinuate, sinuous. Having a strongly wavy margin.

sinus.   The indentation or space between two lobes or divisions; e.g. as in a lobed leaf, or between the semicells of certain desmids.

siphonaceous, siphonous.   A morphological type of growth form which is multinucleate and nonseptate; e.g. as in green algae; coenocytic.

siphonogamous.   Having pollen tubes.

siphonogamy.   A reproductive process in which a pollen tube carries the sperm cells to the egg; this is typical of most seed plants, but with certain exceptions, e.g. cycads and *Ginkgo*. Compare zooidogamy.

siphonostele.   A stelar arrangement in which the vascular tissue forms a cylinder around a central pith. *See* amphiphloic siphonostele and ectophloic siphonostele.

siphonous line.   A proposed evolutionary sequence of siphonaceous green algae (Chlorophyta).

sirenin.   A hormone produced by the female gametes of some fungi (e.g. *Allomyces*) which attracts male gametes.

sister chromatids.   The two chromatids produced by the replication of a single chromosome.

slide.   A sheet of glass on which a specimen is placed (mounted) for examination under a microscope.

slime.   *See* P-protein.

slime body.   An aggregation of P-protein frequently found in young sieve elements; disperses later as slime.

slime mold.   A common name usually referring to a fungus in the class Myxomycetes, the true slime molds. *See* cellular slime mold.

slip.   A horticultural term for the offshoots of certain plants. *See* offset.

slipping (of bark).   The separation of bark from the wood; results from the breakage of weak radial walls of cambial cells that are undergoing rapid growth during the spring.

slope.   The incline of the surface of the ground.

slug.   The pseudoplasmodium of cellular slime molds, (Acrasiomycetes). (*See* fig. S-8).

smooth.   Lacking roughness or pubescence. *See* glabrous.

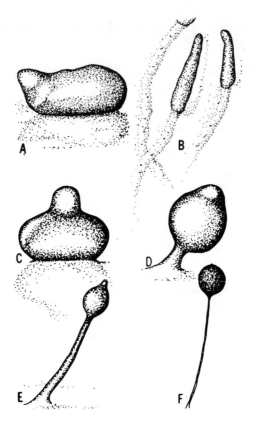

Fig. S-8. Slug: Pseudoplasmodial development A,C,D,E,F. B. trail left as slug moves.

smooth ER.   Endoplasmic reticulum which does not have ribosomes on its surface. Contrast rough ER.

smut.   1. The mass of dark powdery chlamydospores formed in sori of smut fungi (Ustilaginales). **2.** A fungal disease caused by a member of the Ustilaginales. May be one of two types: *a.* covered, in which the spore mass stays within the sorus, frequently until after the dispersal of the sorus, and *b.* loose, in which the mature spores are an uncovered mass of powder disseminated by wind and rain.

soboliferous.   Bearing suckers; producing shoots from the ground, forming clumps.

sod.   The top few inches of soil that is filled with grass roots.

sod grasses.   Grasses that spread by stolons or rhizomes forming a dense mat of grass. Compare bunch grasses.

softwood.   A common term for the wood of conifers; characterized by lack of vessels. Softwood is actually a misnomer since some conifers have hard wood.

soilage.   Forage grown to be cut when green and used directly as livestock feed.

soil profile.   The succession of soil zones seen in a vertical cut through a soil.

soil texture.   The relative proportions of sand, silt, and clay in a soil.

sole.   The basal portion of a carpel.

soleaform.   Slipper-shaped, or shaped somewhat like an hourglass.

solenostele.   An amphiphloic siphonostele; i.e. having phloem both internal and external to the xylem; characteristic of rhizomes.

solitary.   Borne singly or alone.

solitary pore.   A pore surrounded by cells other than vessel members.

soluble RNA.   *See* transfer RNA.

solute.   A dissolved substance.

solute potential.   The relative free energy of water in a cell as it is affected by the amount of dissolved solutes. Expressed in negative atmospheres. *Syn.* osmotic potential.

solution.   Usually a liquid, in which molecules of a solute are evenly dispersed among molecules of a solvent; a homogenous mixture of two or more substances.

solvent.   Any substance, usually a liquid, in which other substances are dissolved.

soma, *pl.* **somata.**   The vegetative body of an organism (plant).

somatic.   Refers to the vegetative phase of plants; i.e. the usually diploid cells which make up the nonreproductive, vegetative tissue of multicellular plants and animals. *Syn.* somatic body, somatic cells, somatic tissue.

somatic apospory.   Development of an embryo sac from nucellar cells other than from a sporogenous cell.

somatic cells.   *See* somatic.

somatic doubling.   The doubling of a diploid chromosome set. Can be induced experimentally by applying colchicine in a lanolin paste to somatic tissues which are undergoing mitosis.

somatic mutation.   A mutation which takes place in vegetative tissues.

somatic tissue.   *See* somatic.

somatogamy.   The fusion of somatic cells (e.g. vegetative hyphae), during plasmogamy.

sordid.   Having a dull, dingy or dirty hue; appearing dirty as opposed to white.

soredium, *pl.* **soredia.**   In lichens, a specialized asexual reproductive structure consisting of a few algal cells surrounded by fungal hyphae. When separated from the thallus, they may form a new lichen organism.

sorocarp.   An undifferentiated fruiting structure of the cellular slime molds (Acrasiomycetes).

sorophore.   **1.** A stalk that supports the sorus in cellular slime molds (Acrasiomycetes). **2.** A gelatinized portion of the sporocarp wall of water ferns that bears sori, e.g. *Marsilea.* After the sporocarp imbibes water, the sorophore greatly elongates, thereby exposing the sori to the outside. (*See* fig. S-9).

sorosis.   A fleshy multiple fruit derived from the coalesced ovaries of several flowers on a common axis.

sorus, *pl.* **sori.**   A cluster of spores and/or sporangia, as in ferns. (*See* fig. S-9).

spadix.   A spike borne on a succulent axis enveloped by a spathe. Characteristic of the aroids (Araceae). (*See* fig. S-10).

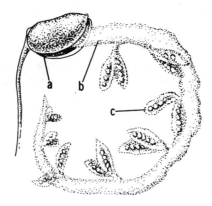

Fig. S-9. Sorophore: a. sporocarp, b. sorophore, c. sorus.

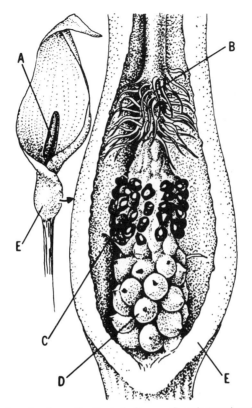

Fig. S-10. Spathe and spadix: A. spadix, B. sterile flowers, C. male (staminate) flowers, D. female (pistillate) flowers, E. spathe.

sparse.   Scattered.

spathaceous.   Spathelike.

spathe.   A large, usually solitary, often showy, sheathing bract which encloses a spadix, as in the aroids (Araceae). *See* cymba. (*See* fig. S-10).

spathe valves.   The herbaceous or scarious bract or bracts which subtend a structure, as an inflorescence or flower; they generally subtend the structure in the bud. Found in certain monocotyledons.

spatulate.   Spoon or spatula-shaped. *Syn.* spathulate.

specialized.   Derived or modified; in organisms refers to specialized adaptations; in cells refers to having a specialized function.

speciation. The process by which a new species comes into existence.

species, *pl.* **species.** A type of organism; species are designated by a binomial consisting of the generic name and the species epithet. Species are variously defined, but are usually considered to be a group of organisms (population) that actually (or potentially) interbreed, and are reproductively isolated from all other such groups.

species-abundance curve. A frequency curve which plots the number of species (on the ordinate) against numbers of individuals per species (on the abscissa).

species-area curve. A curve which shows that species diversity (i.e. number of species) increases with progressively larger sample areas.

species diversity. The number of species found in a given area; e.g. as in a community.

species nova. Indicates a new species.

species packing. The process of increasing the numbers of species in a given area. Involves various changes in the ecological requirements of resident or immigrant species; also there is a limit to the closeness of packing.

species-specific. Characteristic of, and limited to, a particular species.

species transformation. The change of species A into species B with the passage of time; does not increase the number of species.

specific. Of or pertaining to a species.

specific activity. The ratio of radioactive to nonradioactive atoms or molecules of a single type. Sometimes given as the number of atoms of a radioisotope per million atoms of the stable element. Also expressed in curies per mole.

specific epithet. The latin appellative given to a species; with the generic name forms the binomial. Sometimes just referred to as the epithet.

specific-gravity separator. A seed-cleaning device which separates foreign debris and weed seed from desirable seed by differences in seed density.

specific heat. The amount of heat (in calories) required to raise the temperature of one gram of substance one degree centigrade; the specific heat of water is one calorie.

specificity. Uniqueness, as in specific enzymes being required for a given reaction to occur.

specimen. **1.** An individual plant, representative of a population. **2.** A plant or portion of a plant, prepared for botanical investigation.

specimen screen. A support disc for sections to be examined with an electron microscope.

spectrum. The range of visible colored bands diffracted and arranged in order of their respective wavelengths from longest (red) to shortest (violet) as a result of white light being passed through a prism or other defracting medium. Can also include invisible components at both ends of the spectrum, i.e. infrared and ultraviolet.

speiranthy. Having a twisted flower.

-sperm, sperma-, sperme-, spermato-. A suffix or prefix meaning seed.

sperm. A male gamete which is usually motile and smaller than the female gamete. *Syn.* sperm nucleus. (*See* fig. A-9).

spermagonium, *pl.* **spermagonia.** In the rust fungi, the structure that produces spermatia. *Syn.* spermogonium.

spermatiophore. A specialized hypha that produces spermatia.

spermatium, *pl.* **spermatia.** A small nonmotile male gamete found in some sac fungi (Ascomycetes), rusts (Uredinales, Basidiomycetes) and red algae (Rhodophyta).

spermatization. Plasmogamy by the union of a spermatium with a receptive structure.

spermatocyte. Sporocytes (sperm "mother cells") which give rise by meiosis to sperm cells.

spermatogenous cell. **1.** One of two diploid cells, derived from the generative cell of gymnosperm pollen grains that divides mitotically to produce two sperm cells. This term is synonymous with, and perhaps preferable to the term body cell. **2.** A cell which gives rise to sperms. *Syn.* androgenous cell.

spermatophyte. A seed plant.

spermatozoid. A motile, ciliated male gamete.

sperm nucleus. *See* sperm.

spermogonium. *See* spermagonium.

sphaerocyst. Spherical cells found in the trama of some fungi in the family Russulaceae.

S phase, S period. The phase in a cell cycle during which DNA synthesis occurs during interphase; i.e. when the chromosomes are duplicated.

sphenoid.   Cuneate; wedge-shaped.

sphenopsid.   Any member of a group of vascular plants characterized by whorled leaves and terminal strobili; e.g. *Equisetum* and related species, extant and extinct.

spherical, spheroidal.   Having the form of a sphere.

spherosome.   Small spherical bodies (0.7 to 0.9 micrometers in diameter), found in the cytoplasm of plant cells, especially those which are metabolically active. They contain lipids and apparently are centers of lipid synthesis and accumulation.

spherule.   Multinucleate segments of the sclerotia of slime molds (Myxomycetes); they give rise to new plasmodia when adequate moisture becomes available.

sphingophilous, sphinophily.   Pollination by hawk moths and other nocturnal moths.

spicate.   Having the form of or produced in a spike.

spiciform.   Spikelike.

spiculate, spiculose.   Having a surface covered with fine points, or crystals.

spike.   An unbranched, indeterminate, elongated inflorescence which bears sessile flowers. (*See* fig. P-1).

spikelet.   A small or secondary spike; the characteristic inflorescence of the grasses (Poaceae) and sedges (Cyperaceae).

spikelike.   Resembling a spike.

spindle.   The spindle-shaped aggregation of microtubules involved with chromosome movement during mitosis and meiosis. (*See* fig. S-11).

spindle fibers.   An aggregate of microtubules forming a spindle-shaped complex that extend from the centromeres of the chromosomes to the poles of the spindle or from pole to pole in a dividing cell. (*See* fig. S-11).

spindle-shaped.   Widest in middle and tapering toward either end; fusiform.

spine.   A hard, sharp-pointed structure, usually a modified leaf, portion of a leaf or stipule. Compare thorn.

spinescent.   Spine-tipped, having spines.

spinose, spinous.   Spinelike, or having spines.

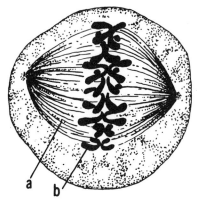

Fig. S-11. Spindle and spindle fibers: a. spindle fiber, b. chromosome.

spinule.   A small spine or spinelike structure.

spinulose.   Having small spines.

spiny.   Covered with spines.

spiral.   Twisted around an axis, like a corkscrew.

spiral cell wall thickening.   *See* helical thickening.

spiricle.   A delicate, coiled, hygroscopic threadlike filament found in the surface cells of certain seeds and achenes which uncoils when moistened; as in *Ruellia.*

spirillum, *pl.* **spirilla.**   A spiral or corkscrew-shaped bacterium. *Syn.* spiril.

spirit collection.   A collection of plants or plant parts preserved in liquid preservatives.

spirochete.   A nonflagellated, spiral bacterium that moves by body flexions.

spongy.   Cellular; spongelike.

spongy parenchyma.   Chloroplast-bearing leaf mesophyll cells which are loosely arranged with conspicuous intercellular spaces. *See* palisade parenchyma. (*See* fig. M-4).

spontaneous generation.   An antiquated idea that living organisms can originate from nonliving matter independent of other living matter. *Syn.* abiogenesis.

sporangiolum, *pl.* **sporangiola.**   A small sporangium without a columella and with a small number of spores. *Syn.* sporangiole.

sporangiophore. **1.** A hyphal stalk that bears a sporangium at its apex; as in some Zygomycetes. (*See* fig. S-12). **2.** The appendages that bear the sporangia in the strobilis of *Equisetum*.

sporangiospore. A spore produced in a sporangium.

sporangium, *pl.* **sporangia.** A hollow, unicellular or multicellular saclike, spore producing structure. (*See* figs. A-9, S-12).

spore. A haploid reproductive cell, produced as a result of mitosis or meiosis, capable of developing into an adult without fusion with another cell; spores are usually unicellular, but may be multicellular, and may undergo mitosis to produce gametes.

spore mother cell. *See* sporocyte.

spore print. The design created as a result of spores falling onto a flat surface (such as a piece of paper) from the cap of a gill fungus (Basidiomycetes).

spore tetrad. The four meiospores produced from a single spore mother cell by meiosis.

sporidium, *pl.* **sporidia.** A basidiospore found in smuts and rusts.

Fig. S-12. Sporangiophore: 1–3: developmental sequence of sporangium in *Philobolus*; A. sporangium, B. subsporangial swelling, C. water droplet, D. sporangiophore, E. trophocyst.

sporiferous. Bearing or producing spores.

sporocarp. A hard, multicellular, nutlike receptacle which contains sporangia in heterosporous ferns, e.g. *Marsilea* (*See* figs. H-5, S-9), and in some fungi, e.g. Ascomycetes and Basidiomycetes. *Syn.* rhizocarp.

sporocladium, *pl.* **sporocladia.** A specialized fertile branch of a sporangiophore which bears merosporangia.

sporocyte. 1. In angiosperms, a diploid cell found in either the anther or the ovary that undergoes meiosis producing four haploid cells (spores) or nuclei. *Syn.* spore mother cell. **2.** A simple sporangia which contains spores.

sporoderm. 1. The wall of a spore. **2.** In angiosperms, the wall of the pollen grain (microspore) composed of two layers: the exine (outer wall) and intine (inner wall).

sporodochium, *pl.* **sporodochia.** A compact, cushionlike stroma covered with conidiophores found in some Fungi Imperfecti.

sporogenesis. The process of producing spores.

sporogenous. A tissue which produces reproductive parts or organs.

sporophore. Any structure that bears spores.

sporophyll. A modified leaf or leaflike structure which bears sporangia; e.g. the stamens and carpels of the angiosperms, fertile fronds of ferns, etc.

sporophyte. The diploid (2n), spore producing generation in the life cycle of a plant which has an alternation of generations.

sporopollenin. The chemical substance which composes the exine (outer wall), of pollen grains and spores; it is a cyclic alcohol which is extremely tough and decay resistant.

sporothallus. A thallus that produces spores, as opposed to one that produces gametes (gametothallus).

sport. A plant or portion of a plant that arises by spontaneous mutation; phenotypically it exceeds the normal limits of variability usually associated with that species. Horticulturally, sports are important sources of new plant material and must usually be maintained asexually (vegetatively); e.g. naval orange, many apple varieties, etc.

sporulate. To produce spores.

spp. The plural abbreviation for species.

spray.   The small branches or branchlets of trees, with their foliage.

spreading.   Diverging outward, almost horizontally.

sprig.   A small shoot or twig.

spring wood.   *See* early wood.

sprout.   **1.** A new shoot of a plant, as the first shoot from a seed. **2.** To germinate.

spumous.   Frothy.

spur.   **1.** A hollow projection of the corolla or calyx, usually functioning as a nectar receptacle. **2.** Lateral stems on the branches of certain woody plants whose growth is restricted, characterized by greatly shortened internodes; they may bear leaves, flowers, and/or fruits. *Syn.* short shoot. (*See* fig. S-13).

squama, *pl.* **squamae.**   A scale, usually derived from a leaf.

squamellate.   Appearing like a small scale; with small scales.

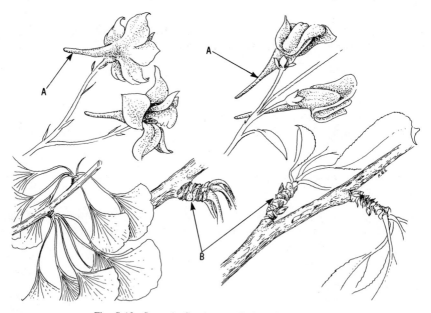

Fig. S-13. Spur: A. floral spurs, B. branch spurs.

squamose, squamaceous, squamate.   Being covered with small scales; scaly.

squamule.   A small, loosely attached lobe in a squamulose lichen.

squamulose.   A lichen growth form similar to foliose but with numerous, small squamules.

square ray cell.   A nearly square ray cell as viewed in a radial section of secondary vascular tissues.

squarrose.   Spreading or recurved at some point above the base; e.g. as in the phyllaries of some composites (Asteraceae) which are sharply curved downward or outward.

squarrulose.   Barely, or minutely squarrose.

sRNA.   *See* transfer RNA.

stabilizing selection.   Selection which favors individuals with average characteristics and eliminates extreme variants. *Syn.* normalizing selection.

stable isotope.   A nonradioactive isotope of an element.

stalk.   The main supporting axis of any organ or structure; a "stem".

stalk cell. *See* sterile cell.

stamen.   In angiosperms, the pollen producing structure in a flower, usually consisting of an anther and a filament. Collective, the stamens are called the androecium.

staminal disc.   A fleshy, elevated cushion of tissue found at the base of an ovary; formed from coalesced staminodia or nectaries.

staminate.   Having only stamens in a flower; a male flower lacking a pistil. (*See* figs. C-15, S-10).

staminiferous.   Bearing stamens.

staminode.   A sterile stamen which does not produce pollen; they are variable in form, petallike and showy in some species. *Syn.* staminodium, *pl.* staminodia.

standard.   **1.** The uppermost petal of a papilionaceous corolla, as in certain members of the pea family (Fabaceae). *Syn.* banner. **2.** The erect or ascending, narrow portion of the perianth of an *Iris* flower. Compare falls.

standard deviation.   A measure of variability in a population of individuals.

standard error. A measure of the variation of a population of means. The standard deviation divided by the square root of the number of observations.

standing crop. *See* biomass.

staphylococcus. A spherical bacterium which occur in irregular, grapelike clusters.

starch. A complex insoluble carbohydrate, composed of many glucose units (monomers), and which is readily degraded enzymatically into these units. It is the most common storage carbohydrate (food) in plants.

starch sheath. Refers to the innermost layer of the cortex when this layer contains an accumulation of starch.

stasad. A plant which lives in stagnant water.

station. The specific site where a plant grows and is collected.

statistical significance. *See* significance, test of.

statistics. The scientific discipline concerned with the analysis of variability in data. Such analysis depends on the application of probability theory.

statocyst. A root cap cell containing statoliths.

statolith. Solid, cellular inclusions, principally amyloplasts (starch grains), that occur in certain root cap cells (statocysts); they perhaps are involved in geotropic response of roots.

statospore. In certain algae (e.g. Chrysophyta and Xanthophyta), a type of resting cell or cyst formed within a cell; upon germination it liberates one or two motile zoospores.

stelar. Referring to a stele.

stele. The portion of the plant body which comprises the vascular system (xylem and phloem) and its associated ground tissue (e.g. pericycle, interfascicular regions and pith). *Syn.* central cylinder.

stellate. Star-shaped hairs; plant trichomes which have radiating branches.

stelliform. Star-shaped.

stellulate. Minutely stellate.

stem. The above ground axis of vascular plants, as well as anatomically similar below ground portions, e.g. rhizomes, bulbs, corms.

steno-. A prefix meaning narrow.

stenoplastic. A genotype with a narrow range of plasticity. *Opp.* euryplastic.

stephanokont. Having an anterior ring of flagella of equal length, as in some green algae (Chlorophyta).

stereoisomers. Molecules which have identical structural formulas but which differ in the spatial arrangement of the atoms or groups of atoms.

stereom, stereome. A collective physiological term for all supporting tissues in a plant; e.g. sclerenchyma and collenchyma.

stereomorphic. Flowers three-dimensional with radial symmetry; with parts many or reduced, and usually regular, e.g. *Narcissus*.

sterigma, *pl.* **sterigmata.** A small, slender stalk which bears conidia or basidiospores.

sterile. Infertile; lacking functional reproductive structures or organs.

sterile cell. One of two cells produced by mitotic division of the generative cell in gymnosperm pollen grains; it is a sterile spermatogenous cell which eventually degenerates. *Syn.* stalk cell.

sternotribic. Flowers with stamens and/or stigmas arranged so as to strike insect visitors on the ventral portion of the thorax. Compare nototribic.

stigma, *pl.* **stigmas.** **1.** The portion of a carpel upon which pollen germinates; usually a glandular surface. (*See* fig. F-3). **2.** The eyespot of algae.

stigmatic. Referring to the stigma.

stigmatoid tissue. *See* transmitting tissue.

stilt roots. *See* prop roots.

stinging hair. A hollow plant trichome which secretes an irritating fluid, as in the nettle family (Urticaceae).

stipe. A supporting stalk, such as the stalk of a pistil, a gill fungus, or the petiole of a fern leaf (*See* fig. H-4).

stipel. A stipulelike appendage at the base of a petiolule of a leaflet.

stipellate. Having stipels.

stipitate.   Borne on a stipe or stalk.

stipular.   Having or pertaining to stipules.

stipular scar.   The mark left on a stem after the detachment of a stipule.

stipulate.   Having stipules.

stipule.   A small structure or appendage found at the base of some leaf petioles; usually present in pairs; they are morphologically variable and appear as scales, spines, glands or leaflike structures. (*See* fig. S-14).

stock. **1.** That part of the stem and associated root system onto which is grafted a scion. **2.** An artificial breeding group.

stolon. **1.** An aerial stem that grows horizontally along the ground often forming adventitious roots at the nodes, e.g. a strawberry runner. **2.** An aerial hypha which usually bears rhizoids and sporangiophores at the point of contact with the substrate, e.g. in breadmold, *Rhizopus*.

stoloniferous.   Producing or bearing stolons.

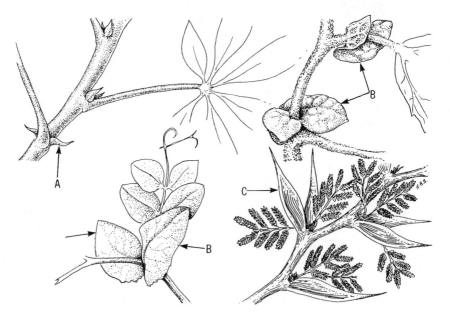

Fig. S-14. Stipules: A. stipule, B. foliar stipules, C. stipular thorn.

stoma, *pl.* **stomata.** A minute pore in the epidermis of leaves or stems which functions in gaseous exchange between a plant and its external environment. Stoma are bordered by guard cells which function in the regulation of the size of the opening; stoma plus guard cells are referred to as the stomatal apparatus. Also called a stomate. (*See* figs. H-6, M-4, S-16).

stomatal apparatus. A collective term for the stoma and any subsidiary cells. *Syn.* stomatal complex.

stomatal crypt. A depression in the leaf epidermis which bears stomata.

stomium. **1.** The lip cell region where dehiscence occurs in fern sporangia. **2.** The fissure or pore through which pollen dehiscence occurs in an anther locule. (*See* fig. T-1).

stone. The hard endocarp of a drupe or drupelet.

stone cell. *See* brachysclereid.

stone fruit. A drupe or pyrene; e.g. plum, peach or apricot.

storage leaf. A succulent, fleshy leaf.

storied cambium. A layered vascular cambium in which the fusiform initials and rays are arranged in horizontal tiers on tangential surfaces. *Syn.* stratified cambium. (*See* fig. S-15).

storied cork. Protective tissue found in monocotyledons which is made up of suberized cells occurring in a series of many-celled radial files.

storied wood. A layered wood in which the axial cells and rays are arranged in horizontal tiers on tangential surfaces. *Syn.* stratified wood.

strain. A group of similar individuals within a taxonomic variety.

stramineous. Straw-colored.

strap. The ligule of a ray floret in the Asteraceae.

strap-shaped. Ligulate shaped; often in reference to the ray floret of composites (Asteraceae).

strasburger cells. *See* albuminous cells.

stratification. **1.** A method of overcoming embryo dormancy of seeds; dormant seeds are placed in a moist medium (e.g. soil or sand) and exposed to either cold or warm

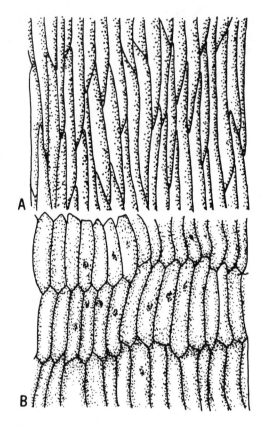

Fig. S-15. Storied cambium: A. nonstoried cambium, B. storied cambium.

temperatures, depending on the required treatment. **2.** Referring to the vertical layering of the organisms within a community.

stratified cambium.   *See* storied cambium.

stratified wood.   *See* storied wood.

streptococcus.   A spherical bacterium which divides in such a way that chains of cells are produced.

stria, *pl.* **striae.**   Narrow lines, grooves, streaks, channels or punctae that are arranged more or less parallel.

striate.   Having fine longitudinal striae arranged more or less parallel.

striate venation.   *See* parallel venation.

strict.   Stiff, rigid, erect; standing upright.

strigillose, strigulose.   Minutely strigose.

strigose.   Having straight, sharp, stiff, appressed hairs frequently with a bulbous base.

strobile.   A cone.

strobilus, *pl.* **strobili.**   A cone; an axis with short internodes which bears an aggregation of sporophylls (spore-bearing appendages), or ovule-bearing scales.

stroma, *pl.* **stromata.**   **1.** The ground substance of plastids. **2.** A mass of vegetative fungal hyphae which may or may not include host tissue; spores may be produced in or on the stroma.

strombiform, strombuliform.   Snail-shaped, usually elongate.

strombus.   A spirally coiled legume; as in bur clover, *Medicago.*

strophiolate.   Having an elongate aril or strophiole in the hilum region.

strophiole.   Outgrowths which occur on the funiculus (raphe) and/or seeds of certain plants following fertilization; e.g. found in *Euonymous* and *Acacia* spp.

structural gene.   A gene in an operon which controls or influences a particular characteristic of an organism by specifying the kinds of amino acids in a polypeptide chain. More or less synonymous with Mendelian gene.

strumose.   Having cushionlike swellings; bullate.

stylar.   Pertaining to the style.

style.   The slender, stalklike extension of a pistil which extends from the ovary to the stigma; it is the tissue through which the pollen tube grows. (*See* fig. I-4).

stylode.   A term proposed to replace the term stylar branch.

styloid.   An elongated crystal with pointed or square ends.

stylopodium.   A disklike expansion at the base of the style, as in the umbel family (Apiaceae).

stylospore.   Globose, detachable, unicellular spores formed on elongated sporangiophores; as in certain Zygomycetes (*Mortierella*).

sub-. A prefix meaning under, below or almost.

suberin. A fatty substance found in the cell walls of cork tissue and in the Casparian strip of the endodermis.

suberization. 1. The impregnation of a cell wall with suberin. 2. The process whereby cut surfaces of tuberous roots heal and develop cork tissue.

suberous. Corky.

subfamily. A taxonomic group of genera within a family.

subfossils. Nonpetrified, more or less carbonized plant remains from Quaternary deposits.

subglabrate, subglabrous. Nearly glabrous.

subiculum, *pl.* **subicula.** A loose hyphal mat upon which sporocarps are borne.

subirrigation. The application of water from beneath the surface of the soil. Water is drawn upward through the soil by capillary action.

submarginal. Close to the margin.

submarginal initial. Cellular components of the marginal meristem of a leaf which appear to contribute cells to the interior tissue of a leaf blade.

submersed. Under water; submerged.

submetacentric chromosome. A chromosome which has a centromere located nearer one end than the other. Chromosome appears J-shaped at anaphase.

suboperculate. An ascus with a thick apical ring capped by a plug or hinged operculum.

subpetiolar. Under the petiole and often enveloped by it; used in reference to buds, as in sycamore trees (*Platanus*).

subshrub. A perennial with stems that are woody at their base; a suffrutescent perennial.

subsidiary cell. A morphologically distinct epidermal cell which is associated with the guard cells. *Syn.* accessory cell. (*See* fig. S-16).

subspecies. A taxonomic subdivision of a species, usually defined as a geographical race.

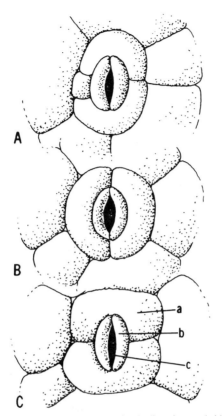

Fig. S-16. Subsidiary cells: A. anisocytic, B. paracytic, C. diacytic; a. subsidiary cell, b. guard cell, c. stoma.

substrate.   1. The surface or medium on or in which an organism is living and from which it gets its nourishment. **2.** The substance acted on by an enzyme.

subtend.   Occurring immediately below and close to, as a bract subtending a flower.

subterranean.   Occurring underground; below the surface.

subtidal.   Referring to that part of the sea floor which is never exposed, even at the lowest, low-tide.

subtropical.   Almost tropical; referring to geographical areas which occur near the tropical regions of the world and which have nearly tropical environments.

subulate.   Awl-shaped, tapering from base to apex.

succession.   In plant ecology, the orderly sequence of changes in a plant community during the development of vegetation in any area. Includes all changes which take place from the initial colonization of a previously unoccupied geographical area through the maturation of that vegetation. (*See* fig. S-17).

succubous.   Having leaves inserted so that the upper part or margin of a leaf is covered by the lower margin of the leaf directly above it; as in liverworts (Jungermanniales). Compare incubous.

succulent.   1. A plant which accumulates water in fleshy, water-storing stems, leaves, or roots. **2.** Juicy, fleshy, in reference to texture or appearance.

sucker.   A shoot which develops from roots or underground stems. *See* offset.

sucrase.   An enzyme that hydrolyzes sucrose into glucose and fructose. *Syn.* invertase.

sucrose.   Common table sugar; a disaccharide (glucose plus fructose) found in many plants. It is the primary form in which sugar, produced as a result of photosynthesis, is translocated in a plant. Commercially derived primarily from sugar cane (*Saccharum*) and sugar beets (*Beta*).

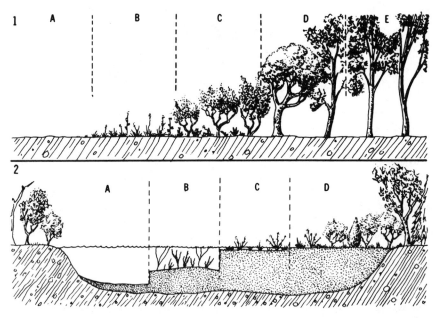

Fig. S-17. Succession: 1: A. barren field, B. grasses and broadleaf herbs, C. herb-shrub cover, D. oak-hickory forest, E. beech-maple forest. 2: A. newly formed pond, B. submerged rooted plants, C. emergent plants, D. beginning of forest succession.

suffrutescent.   A perennial plant which is slightly woody only at the base.

suffruticose.   A degree above suffrutescent in amount of woodiness; i.e. with permanent woody stems extending several inches above the ground level.

sulcate.   Having longitudinal grooves, furrows or channels.

sulcus, *pl.* **sulci.**   A small longitudinal groove, furrow or channel, as in some pollen grains. Also, the longitudinal groove characteristic of the cell walls of dinoflagellates (Pyrrophyta) which frequently contains a flagellum. (*See* fig. D-4).

sulfureous.   Sulfur-colored.

summer annual.   A plant which germinates in the late spring or early summer, flowers and fruits in the late summer or early fall, and then dies.

summer wood.   *See* late wood.

super-.   A prefix meaning above.

superficial.   On or near the surface.

supergene.   A gene complex which is more or less permanently linked and is transmitted as a unit.

superior ovary.   *See* hypogynous. (*See* fig. E-4).

supernatant.   The fluid found above the precipitate in a centrifuge tube following centrifugation of a suspension.

supernumerary cambium layer.   A vascular cambium which originates in the phloem or pericycle and is removed from the normal zone of vascular cambium activity. Characteristic of some plants exhibiting anomalous secondary growth.

supernumerary chromosome.   *See* B chromosome.

superposed bud.   A lateral bud positioned above an axillary bud.

superspecies.   A natural grouping of closely related species, which are often difficult to distinguish morphologically.

supervital mutation.   Any mutation which increases the viability of the individual that bears it, giving that individual greater viability than the wild type.

supine.   Prostrate, with parts oriented upward.

supporting cell.   A specialized cell which bears the carpogonial branch in some red algae, e.g. *Polysiphonia* (Rhodophyta).

supporting tissue.   A general term for any tissue that adds strength to the plant body; composed of cells with cell wall thickenings, such as collenchyma or sclerenchyma cells. *Syn.* mechanical tissue.

suppression.   A restoration of lost genetic function as a result of a suppressor mutation.

suppressor genes.   *See* inhibitor genes.

suppressor mutation.   A change which reverses the effect of a previous primary mutation. Normally occurs at a genetic locus removed from the primary mutation.

supra-.   A prefix meaning above.

suprafoliar.   On the stem above the leaves.

surcarpous.   With fruits borne at or on the soil surface.

surculose.   Producing suckers.

surcurrent.   Having laminar extensions beginning at the leaf base and continuing up the stem.

surfactant.   A monomolecular compound used to prevent evaporation of water from the surfaces of large bodies of water, e.g. lakes.

survival value.   An estimate of the ability of an individual phenotype to contribute offspring to future generations.

susceptibility.   Capable of being infected; not able to resist some foreign agent such as a pathogen or a drug.

suspension.   A heterogeneous dispersion in which one particulate substance is dispersed through a fluid, but not dissolved in it. Particles will settle out of suspension as a result of gravitational forces.

suspensor.   **1.** In many vascular plants, a structure in the embryo which pushes the terminal part of the embryo into the endosperm. (*See* fig. S-4). **2.** In the zygomycete fungi, a hyphal cell which supports a gametangium or zygospore. (*See* fig. Z-1).

suture.   The line resulting from the fusion of contiguous parts; the seam at which dehiscence or splitting occurs, as in a fruit.

S values.   *See* sedimentation coefficient.

Svedberg units.   *See* sedimentation coefficient.

swarm cell.   A motile, flagellated cell which acts as an isogamete, as in the slime molds (Myxomycetes).

syconium.   A hollow, multiple fruit in which flowers and ultimately achenes are borne, on the inside of a receptacle or peduncle; e.g. as the fruitlike structure of a fig.

syleptic shoots.   Abnormal shoots that develop from immature lateral buds.

sylva.   In reference to trees.

sym-.   A prefix meaning together.

symbiont.   One of the organisms in a symbiotic association.

symbiosis.   The living together of two or more dissimilar organisms. This association may benefit only one, or all of the symbionts. See commensalism, mutualism, and parasitism. (*See* fig. S-18).

symbiotic.   Referring to symbiosis.

symmetrical.   **1.** In references to flowers, having the same number of parts in each series or whorl. *Syn.* actinomorphic. **2.** Balanced; having or exhibiting symmetry.

symmetry.   The correspondence of parts in size, form and arrangement on opposite sides of a plane, line, or point.

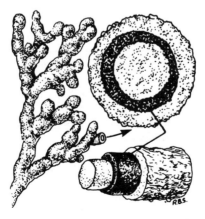

Fig. S-18. Symbiosis: Blue-green algal symbionts in coralloid roots of a cycad.

sympatric, sympatry.   1. Originating in, or occupying the same geographical region. **2.** Species or populations which occur close enough together to be within the range of mutual pollinating vectors. Compare allopatric.

sympatric speciation.   Speciation without geographical isolation.

sympetalous, sympetaly.   Having petals which are partly or completely fused (connate). *Syn.* gamopetalous.

symphogenous.   Having a fruiting body that has arisen from a number of interweaving hyphae.

symphysis.   The fusion of like parts, as petals with petals; coalescence.

symplast.   In a plant, the continuum of protoplasts of many cells, together with the plasmodesmata which connect them.

symplastic growth.   *See* coordinated growth.

sympodial, sympodium.   A branching growth pattern in which the main axis is formed by a series of successive secondary axes, each of which represents one fork of a dichotomy.

sympodial inflorescence.   A determinate inflorescence that appears to be an indeterminate inflorescence, e.g. scorpioid cyme.

syn-.   A prefix meaning together, or united.

synandrium, *pl.* **synandria.**   An androecium in which the anthers are superficially joined but not fused. *Syn.* synandry.

synangium, *pl.* **synangia.**   A compound structure formed by the coalescence (fusion) of sporangia; as in certain ferns.

synantherous.   The simultaneous appearance of leaves and flowers.

synanthesis.   The simultaneous maturation of the stamens and pistils.

synapsis.   The pairing of homologous chromosomes which occurs early in meiosis (prophase I). Crossing-over occurs during synapsis.

synaptinemal complex.   A protein framework found between paired chromosomes during meiosis (prophase I).

syncarpous, syncarpy.   Having united carpels. Contrast apocarpous.

syncolpate.   Descriptive of a pollen grain with anastomosing colpi, fused to form spirals, rings, etc.

syncytium.   *See* coenocytic.

syndetocheilic.   In gymnosperms, a stomatal apparatus in which the subsidiary cells or their precursors are derived from the same protodermal cell as the progenitor of the guard cells.

syndiploidy.   Karyokinesis without cytokinesis. Results in autotetraploid tissue.

synecology.   The division of ecology which deals with the biological and environmental relationships among organisms within a community.

synema, *pl.* synemata.   The column formed by monadelphous stamens, as in the mallow family (Malvaceae).

synergid.   A sterile cell or cells closely associated with the egg at the micropylar end of the mature embryo sac in the ovule of angiosperms. (*See* figs. E-2, O-3).

synergism.   A condition in which the action of two or more separate agents, when used in combination, is greater than the sum of their separate, individual actions.

syngameon.   A specialized evolutionary term referring to plant populations intermediate between the species level and extreme variants of the same species.

syngamodeme.   A unit composed of coenogamodemes which are associated by the ability of at least some of their members to form viable but sterile hybrids under a specified set of conditions. *Syn.* comparium.

syngamy.   The fusion of two gametes (egg and sperm); fertilization.

syngenesious.   Having fused (connate) anthers; especially in reference to members of the sunflower family (Asteraceae) in which the stamens are connate by their anthers forming a cylinder or ring around the style.

synnema, *pl.* synnemata.   A group of closely united, and sometimes fused conidiophores, which bear conidia. *Syn.* coremium, *pl.* coremia. (*See* fig. S-19).

synonym.   A taxonomic name which is rejected in favor of another name, because of misapplication or differences in taxonomic judgement. Rejections are made in accordance with the International Rules of Botanical Nomenclature.

synonymy.   A complete listing of the scientific names applied to a particular taxon. Usually appears with other data, such as date of publication, author of publication, etc.

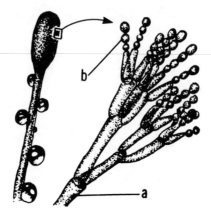

Fig. S-19. Synnema: a. conidiophore, b. conidia.

synoptical collection.   A special collection of specimens kept separate from regular herbarium specimens, and used for teaching or identification.

synovarious.   Having ovaries of adjacent carpels fused but with styles and stigmas separate.

synsepalous, synsepaly.   Having sepals which are partly or completely fused (connate). *Syn.* gamosepalous.

synstylovarious.   Having ovaries and styles of adjacent carpels fused but with separate stigmas.

synthesis.   The production of a more complex compound from simpler substances.

syntropous.   Having the radicle pointing toward hilum.

syntype.   One of several specimens from which a species is described when the holotype was not designated by the author. Also, one of several specimens simultaneously designated as types.

synusia.   A group of plants of the same life form occurring together in the same habitat; the synusia may be composed of the same or unrelated species.

systematics.   The scientific investigation of organisms in regard to their natural relationships, including the description, naming and classifying of these organisms.

systemic.   Of or affecting the entire organism; as a disease in which a single infection leads to its spreading throughout the organism. Also, in reference to various chemicals (especially pesticides) that act through the vascular system of a plant.

T

tabular. Having the form of a tablet or slab.

tactile. Sensitive to touch.

taiga. A conifer forested region of the northern hemisphere.

tail. The basal portion of a bacteriophage. (*See* fig. B-1).

tailed. Having a taillike appendage, a caudal appendage.

talus. Rock fragments derived from the weathering of a cliff or slope and accumulating at its base.

tandem duplication. A chromosomal aberration in which two identical chromosomal segments lie beside one another. The sequence of genes is identical in each segment.

tangential section. A longitudinal section which is cut at right angles to a radius of cylindrical structures, such as a root or stem. (*See* fig. C-12).

tanniferous cells. Cells which contain tannins throughout their cytoplasm, and/or in their cell wall.

tannin. A heterogeneous group of astringent, phenol derivatives which are widely distributed in plants. Used by man in tanning, dyeing and the manufacture of ink. In plants, they act as antiherbivore agents, and protect against desiccation and decay.

tannin sacs. Tannin-containing vacuoles; droplets of tannin within the cytoplasm of a cell.

tapering. Gradually becoming smaller or diminishing in diameter or width toward one end; not abrupt.

tapetum. A nutritive layer of cells found in a sporangium, which provide nourishment to developing sporocytes and spores; in an anther, the innermost parietal wall layer which is absorbed as the pollen grains mature. (*See* fig. T-1).

taproot. The stout, tapering, primary root of a plant which arises as a direct continuation of the embryonic radicle; characteristic of dicots and gymnosperms. This main vertical root bears smaller, lateral branch rootlets.

tassel. The staminate inflorescence of corn.

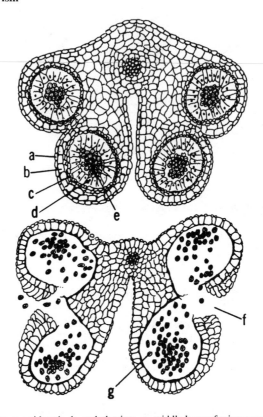

Fig. T-1. Tapetum: a. epidermis, b. endothecium, c. middle layer of microsporangium, d. tapetum, e. microsporocytes, f. stomium, g. pollen grains.

tautomerism. The situation in which two reversible isomeric forms of a molecule exist in equilibrium.

tautonym. An illegitimate binomial having the same generic and specific epithet.

tawny. Dull yellowish-brown; fulvous.

taxis. A movement toward or away from an external stimulus.

taxon, *pl.* **taxa.** Any taxonomic unit into which living organisms are classified, e.g. species, genus or division.

taxonomy. The science of the classification of organisms. More or less synonymous with the term systematics.

tectonic.   Used in reference to the forces and processes of earth movements.

tectum.   In certain pollen grains, the outermost closed layer of the sexine, formed by union of the heads of the bacula.

teichode.   Bundles of specific interfibrillar spaces in the cellulosic wall of epidermal plant cells; appear in cross section as thread-shaped, ribbonlike or conelike structures. Replaces the term ectodesma.

teleology.   The explanation of natural phenomena on the basis of a design or purpose.

teleutospore.   *See* teliospore.

teliospore.   In some rust or smut fungi, a thick-walled spore in which karyogamy and meiosis take place; it gives rise to a short tube (basidium) which bears basidiospores. *Syn.* teleutospore.

telium, *pl.* **telia.**   In some rust fungi, the structure in which teliospores are produced.

telocentric chromosome.   A chromosome with a terminal centromere.

telome.   According to the Telome theory, one of the ultimate distal fertile or sterile branches of a dichotomously branched axis. A hypothetical unit of morphological organization in a primitive vascular plant.

Telome theory.   A theory that regards the primitive vascular plant as consisting of a dichotomously branched axis; one dichotomy is reduced by "overtopping", flattened by "planation" and expanded by "webbing". From such telomes, the diverse types of megaphyllous leaves and sporophylls of vascular plants have evolved.

telophase.   The last stage of mitosis and the first and second division of meiosis, during which the chromosomes are reorganized into two new nuclei.

temperate phage.   A nonvirulent bacterial virus which infects but rarely causes lysis of its host bacterium.

template.   A pattern or mold used to form an accurate negative or complement. In biology, a strand of DNA serves as a template for the formation of a complementary strand of DNA or mRNA.

tendril.   A long, slender, coiling, modified leaf (or rarely stem), by which a climbing plant attaches to its support. (*See* fig. T-2).

tension wood.   The reaction wood of dicotyledons which is found on the upper sides of leaning or crooked branches and stems; it is characterized by a lack of lignification, reduction in width and number of vessels, and by fibers with gelatinous layers (composed mostly of cellulose). Compare compression wood.

tentacular, tentacle.   A sensitive glandular hair, as found in some insectivorous leaves, e.g. *Drosera*.

tenui-.   A prefix meaning thin or slender.

tenuinucellate ovule.   One of two general types of nucellar organization found in angiosperms; the megasporocyte arises directly from a hypodermal cell located at the apical region of a nucellus which consists of single epidermal layer. Compare crassinucellate.

tepal.   A perianth member or segment; term used for perianth parts undifferentiated into distinct sepals and petals.

teratology, teratological.   The scientific study of biological monstrosities and malformations; abnormal.

terete.   More or less circular in cross section; cylindrical and elongate.

terminal.   Found at the tip, apex or distal end.

terminal apotracheal parenchyma.   *See* boundary apotracheal parenchyma.

terminal bud.   A bud found at the apex of a stem axis.

terminalization.   The slipping of a chiasma laterally toward the ends of the chromatids after crossing-over has occurred.

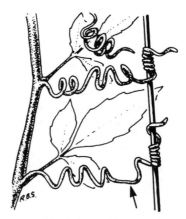

Fig. T-2. Tendril.

termination codon.   A codon that terminates the growth of a polypeptide chain.

ternary.   Trimerous; found in threes.

ternate.   Arranged in threes.

terrestrial.   Growing on land.

tertiary.   1. Third rank. 2. Branches of secondary roots.

tesselate.   Having a checkered pattern, e.g. with depressions.

tesserae.   Functionally different areas of endoplasmic reticulum, each bearing a characteristic set of enzymes.

testa.   The outer seed coat, which is derived from the integuments of the ovule.

test cross.   A cross between an unknown genotype which has at least one dominant gene with an individual homozygous for the recessive genes in question. Used to determine whether the dominant gene is homozygous or heterozygous.

tester strain.   A homozygous recessive genotype used in a test cross.

tetra-.   A prefix meaning four.

tetracoccus.   Spherical bacteria which occur in groups of four.

tetracyclic.   Four-whorled.

tetracytic stomata.   A stoma with four subsidiary cells, two of which are lateral and two terminal. Characteristic of many monocotyledons.

tetrad.   A group of four spores derived from a spore mother cell as a result of meiosis. (*See* fig. T-7).

tetrad analysis.   The analysis of recombination (crossing-over) by examination of all tetrads which have arisen from meiosis in a single primary spore mother cell. Such an analysis is feasible only if the meiotic products are held together, e.g. as in an ascus sac of some Ascomycetes.

tetradynamous.   Having four long and two short stamens; characteristic of the flowers of the mustard family (Brassicaceae).

tetragonal, tetragonous.   Four-angled.

tetrahedral.   Four-sided, as a pyramid.

tetramerous.   Having four members in a whorl.

tetrandrous.   Having four stamens.

tetrapetalous.   Having four petals.

tetraploid.   An organism which has twice the usual, diploid number of chromosomes (4n).

tetrapolarity.   A condition of sexual compatibility in some Basidiomycetes in which each of the four basidiospores of a basidium is of a different strain.

tetrarch.   The primary xylem of a root which has four protoxylem strands or poles.

tetrasomic.   An aneuploid in which one chromosome is represented four times (2n + 2).

tetrasporangium, *pl.* **tetrasporangia.**   A sporangium found in some red algae (Rhodophyta), which gives rise to four spores (tetraspores) after meiosis. (*See* fig. T-3).

tetraspore.   In some red algae (Rhodophyta), the four spores formed by meiosis in a tetrasporangium.

tetrasporic embryo sac.   One of three major types of embryo sac development in angiosperms, in which the embryo sac is derived from four megaspores.

tetrasporine line.   A proposed evolutionary sequence in the green algae (Chlorophyta) which begins with a unicellular prototype and culminates in complex filamentous growth form. This line is considered a precursor to the vascular plants.

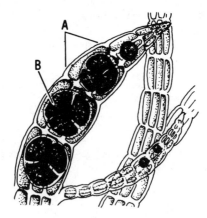

Fig. T-3. Tetrasporangium: A. tetrasporangium, B. tetraspore.

tetrasporophyte. In some red algae (Rhodophyta), a diploid plant that produces tetrasporangia.

tetrastichous. Having structures in four vertical rows.

tetrazolium test. A biochemical method of determining the viability of dormant and nondormant seeds; when seeds are soaked in TTC (2,3,5 triphenyltetrazolium chloride) solution, living tissue stains red, while nonliving tissue remains uncolored.

thalamus. The receptacle of a flower. Floral parts (sepals, petals, stamens and pistils) borne on the receptacle are separate from one another.

thalloid. Resembling or shaped like a thallus.

thallophyte. A term previously used to designate nonvascular plants, e.g. algae and fungi.

thallus, *pl.* thalli. A single plant body which lacks a differentiated root, stem or leaf, e.g. algae and fungi.

theca. A pollen sac or locule of an anther.

thelytoky. A rare type of parthenogenesis in which unfertilized eggs yield females. Males may be absent entirely.

theory. A generalization having predictive value. It is based on considerable observation and experimentation which supports a hypothesis.

thermocline. A demarcation separating an upper layer of warm water from a lower layer of colder water.

thermodynamics, laws of. The first law states that energy can be transformed from one form to another but cannot be created or destroyed. The second law states that the entropy or the degree of randomness or disorder tends to increase in a system.

thermonasty, thermotropism. A growth or nastic movement in response to a change in temperature.

thermoperiodicity, thermoperiodism. Refers to the effects of cyclical alternation of temperature between night and day periods, upon the reactions of plants.

thermophilic. Heat-loving; refers to organisms, such as some bacteria and blue-greens which grow at temperatures between 45° and 70° C.

thermophyllous. Producing leaves in the summer.

therophyte.   An annual plant.

thigmonasty, thigmotropism.   A growth or nastic movement in response to a mechanical stimulis (i.e. touch); e.g. as with tendrils, Venus flytrap leaves, etc.

thorn.   A hard, sharp-pointed, modified branch. Compare spine.

throat.   An opening or orifice in an expanded gamopetalous corolla or gamosepalous calyx, or the somewhat expanded portion between tube proper and limb.

thrum flower.   In heterostylous plants, the flower type having the shorter style. Compare pin flower. (*See* fig. D-5).

thylakoid.   A saclike membranous structural unit of a higher plant chloroplast. Stacks of thylakoids form the grana. (*See* fig. C-6).

thyrse, thyrsus.   A panicle-like inflorescence which has one main indeterminate axis and many lateral axes which are determinate, as in lilac.

thyrsoid.   Resembling a thyrse.

tiller.   A shoot growing from the base of a grass (Poaceae) stem.

timber line.   The upper limit of tree vegetation at high altitudes or latitudes.

tinsel flagellum.   A flagellum which bears numerous short filaments (mastigonemes) along its length. See mastigonemes. Compare whiplash flagellum. (*See* fig. W-1).

tissue.   A group of cells similar in origin, organized into a structural and functional unit.

tissue culture.   **1.** The technique of cultivating cells, tissues or organs in a sterile, synthetic media; includes tissues excised from an organism, and also culture of pollen, seeds or spores. **2.** The mass of more or less undifferentiated callus tissue growing on an artificial medium.

tissue system.   A tissue or group of tissues in a plant organized into a structural or functional unit. Commonly three tissue systems are recognized: dermal, vascular and fundamental (or ground).

toadstool.   A common name for a poisonous mushroom.

tomentose.   Covered with dense, matted, woolly hairs.

tomentulose.   Covered with relatively short, fine woolly hairs.

tomentum.   A covering of woolly, densely matted hairs; wool.

tonoplast.   A one-layered, cytoplasmic membrane which surrounds a vacuole in plant cells. *Syn.* vacuolar membrane.

tooth.   Any small, pointed, toothlike process or marginal lobe.

toothed.   Bearing teeth, dentate; with minor projections along the margin.

topiary.   The pruning and training of trees and shrubs to resemble a variety of shapes, especially animals; also, the trees or shrubs resulting from such treatment. (*See* fig. T-4).

topocline.   A geographic variational trend which is not necessarily correlated with an ecological gradient.

topodeme.   A group of individuals of a taxon which occur in a specified geographical area.

topotype.   A specimen collected at the original type locality of a particular species.

torose.   Cylindrical with constrictions at more or less regular intervals.

tortuous.   Irregularly twisted or twining.

torulose.   Minutely torose.

torus, *pl.* **tori.**   **1.** The receptacle of a flower. **2.** The central thickened part of the pit membrane in a bordered pit of conifers and some other gymnosperms. (*See* fig. B-6).

Fig. T-4. Topiary.

totipotency.   The capacity of a cell or cells to grow and develop into an entire organism.

totipotent.   A cell capable of growing and developing into an entire organism. Many cells of a mature plant are totipotent, but those which lack cytoplasm and a nucleus are not, e.g. tracheids, fibers, sclereids.

toxin.   A poisonous substance produced by living organisms.

trabecula, *pl.* **trabeculae.**   A cell, row of cells, or structure which partially or completely traverses an intercellular space; cross-barred. (*See* fig. M-6).

trace.   A strand of vascular tissue connecting the stem with a leaf or reproductive organ.

trace element.   *See* micronutrient.

tracer.   *See* radioactive isotope.

tracheary element.   A general term for a water conducting cell in vascular plants, e.g. tracheids or vessel elements. Replaces the term trachea.

tracheid.   An elongated, thick-walled, nonliving conducting and supporting cell found in the xylem of most vascular plants. Such tracheary elements have tapering ends which lack perforations.

tracheophyte.   A plant which has vascular tissue (xylem and phloem); e.g. seed plants.

trachycarpous.   Rough-skinned fruits.

trachyspermous.   Rough-coated seeds.

trailing.   Prostrate on the ground, but not rooting.

trama.   The central portion of the gills of some club fungi (Basidiomycetes); consists of longitudinal hyphae from which the hymenium arises.

transcription.   The formation of messenger RNA on a DNA template.

transduction.   The transfer of genetic material (DNA) between bacteria by a lysogenic bacteriophage.

transect.   A sampling line, or belt extending across a stand of vegetation. Used to quantify the vegetation in a given area.

transection.   *See* transverse section.

transfer cell. A specialized parenchyma cell which apparently functions in the short-distance transfer of solutes. The cell wall of these cells has invaginations which increase the surface of the plasma membrane.

transfer RNA, tRNA. A low molecular weight (small) form of RNA. In the cytoplasm, it attaches to an amino acid and transfers it to the correct site on mRNA associated with a ribosome for polypeptide (protein) synthesis. *Syn.* soluble RNA, and sRNA.

transformation. An heritable genetic modification caused by the incorporation of DNA into one cell from a donor cell.

transfusion tissue. A tissue surrounding or otherwise associated with the vascular bundle in gymnosperm leaves. This tissue is composed of transfusion tracheids and parenchyma cells.

transfusion tracheid. A tracheid in transfusion tissue.

transgressive. The appearance in later generations of genotypes which go beyond the limits of variation in the phenotypes of the parents in the F_1 generation.

transient polymorphism. A temporary polymorphism found during the time when one adaptive type is being replaced by another.

transition. A mutation caused by the substitution of purine for a purine or a pyrimidine for a pyrimidine.

transition region. The region where the root and shoot meet; shows transitional characteristics between root and shoot, i.e. where the xylem-phloem arrangement changes from radial to collateral. *Syn.* transition zone.

transition zone. **1.** An intercalary meristem between the lamina and stipe in some brown algae (Phaeophyta). **2.** An ecotone.

translation. The formation of a protein directed by a specific mRNA molecule which is formed on a single strand of DNA, the genetic information.

translator. The hygroscopically active structure connecting the pollen mass of some milkweeds (Asclepiadaceae) to the corpusculum.

translocation. **1.** The movement of water, minerals or food within plants. **2.** An interchange of chromosomal segments between two or more nonhomologous chromosomes. (*See* fig. T-5).

translucent. Semitransparent.

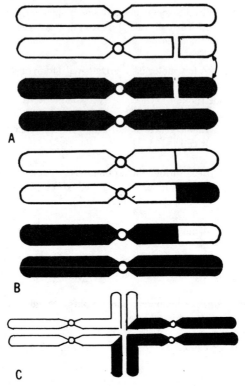

Fig. T-5. Translocation: A. non-sister chromatid exchange, B. resulting change, C. pairing in hetero-zygote.

transmitting tissue. Tissue in the style of angiosperms through which the pollen tubes grow. It is cytologically and physiologically similar to cells of the stigma. *Syn.* stigmatoid tissue.

transmutation. The transformation of one element into another as a result of radioactive decay or by nuclear bombardment.

transparent. Permitting the passage of light rays; capable of being seen through.

transpiration. The loss of water vapor from plant tissues primarily through stomata (stomatal transpiration); also, loss of water through the cuticle (cuticular transpiration) and lenticels (lenticular transpiration).

transpiration ratio. The ratio of water transpired to the dry matter produced by a plant.

transpiration stream. The upward translocation of water in the xylem, from roots to leaves; results from the transpiration of water from the leaves.

transplanting. The transfer of part or all of a living plant from one place to another.

transverse. Across; perpendicular to the long axis.

transverse division. A division perpendicular to the longitudinal axis of a cell or long axis of a plant part.

transversely septate basidium. An elongate basidium divided by cross walls into four cells.

transverse section. A cross section. A section cut at right angles to the longitudinal axis of a plant part. *Syn.* transection.

transversion. In DNA or RNA, a mutation caused by the substitution of a purine for a pyrimidine or vice versa.

trap. A structure which imprisons, e.g. the modified leaf of an insectivorous plant.

trap blossom, trap flower. A flower that imprisons insect visitors until pollination is effected. *Syn.* prison flower.

trap hairs. Trichomes that fill the constrictions of tubular perianths (e.g. *Aristolochia*) or inflorescences (e.g. *Ficus*) to trap pollinators within specially developed chambers; also, the marginal trichomes on various structures of certain insectivorous plants that reinforce the trapping function, as in *Dionaea, Utricularia*.

traumatic. Of or pertaining to a wound.

traumatic resin duct. A resin duct which develops in response to injury.

traumatism. An abnormal growth which results from injury.

tree. A woody perennial plant which usually has a single trunk.

tri-. A prefix meaning three.

triad. Associated in threes.

triadelphous. Having three sets of stamens.

triandrous. Having three stamens.

triangulate. Three-angled.

triarch. Having three protoxylem strands or poles in the primary xylem of a root. (*See* fig. T-6).

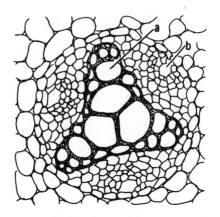

Fig. T-6. Triarch: a. xylem vessel, b. phloem.

tribe.   A taxonomic subdivision of a family.

tricarboxylic acid cycle (TCA cycle).   *See* Krebs cycle.

tricarpellate.   Having three carpels.

trich-, tricho-.   A prefix meaning hair.

trichasium.   A cymose inflorescence having three branches.

trichoblasts.   **1.** A delicate hairlike branch found in some red algae (Rhodophyta). **2.** Protoderm cells that give rise to root hairs or trichomes.

trichocarpous.   Having hairy fruits.

trichocyst.   A cytoplasmic organelle ejected when the organism is disturbed. Found in some yellow-green algae (Chrysophyta). Also called ejectosomes.

trichogyne.   A hairlike extension of the female gametangium which receives the male nucleus, as in red algae (Rhodophyta), and some sac fungi (Ascomycetes) and club fungi (Basidiomycetes).

trichome.   **1.** In vascular plants, an epidermal outgrowth, such as a hair or scale. **2.** In the blue-greens, a single row or chain of cells in a filamentous colony.

trichosclereid.   A branched sclereid with thin hairlike branches extending into intercellular spaces. (*See* fig. S-3).

trichothallic.   Meristematic activity (intercalary growth) occurring from the base of hairlike, uniseriate filaments in some brown algae (Phaeophyta).

trichotomous.   Forking or branching into three divisions.

tricolpate.   Having three grooves. Pollen grains with three furrows or grooves; characteristic of the pollen of dicotyledonous plants.

tricussate.   Having whorls of three leaves each; ternate.

tricyclic.   Having three-whorls.

tridynamous.   Having six stamens in two groups of three each.

trifid.   Three-cleft nearly to the middle.

trifoliate.   Having three leaves.

trifoliolate.   A compound leaf with three leaflets.

trifurcate.   Having three forks or branches.

trigonal.   Three-angled.

trigone.   In liverworts, triangular thickenings in the corners of cell walls.

trigonous.   Three-angled.

trihybrid.   A hybrid heterozygous for three pairs of genes.

trilacunar node.   In a stem, a node which has three leaf gaps associated with one leaf.

trilete.   **1.** Basically tetrahedral, often appearing round or triangular. **2.** Having three scar lines forming a "Y". (*See* fig. T-7).

trilobate.   Having three lobes.

trilocular.   Having three locules.

trimerous.   Having the parts in threes.

trimonoecious.   A species which has staminate, pistillate and perfect flowers on the same plant.

trimorphic.   A characteristic which has three different forms within the same species.

trinomial.   A plant name consisting of the genus, species and subspecies or variety, e.g. *Lycopersicon esculentum* var. *cerasiforme* (cherry tomato).

Fig. T-7. Trilete: A. trilete or Y-shaped scar from tetrahedral tetrad pollen grain association, b. tetrahedral tetrad pollen grain arrangement.

trioecious.   A species which has separate plants that bear either staminate, pistillate or perfect flowers.

triose.   Any three-carbon sugar.

tripalmately compound.   Having three sets of palmately compound leaflets.

tripartite.   Divided into three parts.

tripinnately compound.   Having three sets of pinnately compound leaflets.

triple fusion.   In angiosperms, the fusion of one male gamete with two polar nuclei; (the other male gamete fuses with the egg cell). Results in the formation of the primary endosperm nucleus which is usually triploid (3n).

triplet code.   A code which specifies a given amino acid with a sequence of three nucleotides.

triploid, triploidy.   An organism which has three complete sets of chromosomes per cell (3n).

tripterous.   Having three wings, as on certain fruits or seeds.

triquetrous.   Having three edges, with the faces between them concave.

trisomic.   An aneuploid organism that is otherwise diploid, but has three chromosomes of one type (2n + 1).

tristichous.   Having structures present in three vertical rows.

tristyly.   The condition of a flower in which the anthers are of two different lengths and the styles are of three different lengths (correlated with specific stamen positions).

triternate.   Three times ternate; as leaflets in three sets, each ternately compound.

tritium.   A radioactive isotope of hydrogen, ^{3}H.

trivalent.   An association of three homologous or partially homologous chromosomes during meiosis. Usually results in some gamete sterility.

tRNA.   *See* transfer RNA.

-troph, tropho-.   A suffix or prefix meaning nourish, food or nourishment.

trophic level.   A level of organisms within a food chain, representing a step in the movement of energy through an ecosystem. Trophic levels are depicted in pyramids, with each level containing only 10 percent of the energy below it.

trophocyst.   A yellowish, swollen segment of the mycelium of *Pilobolus* (Zygomycetes), from which sporangiophores develop. (*See* fig. S-12).

tropism.   A movement or growth in response to an external stimulus. The movement or growth may either be towards (positive tropism) or away from (negative tropism) the stimulus. A tropic movement.

true indusium.   An epidermal outgrowth which covers a sorus in some ferns.

trullate.   Trowel-shaped; having its widest axis below the middle and with straight margins.

truncate.   A base or apex which ends abruptly, almost at right angles to the main axis, as if cut or squared off.

trunk.   The main axis, as the main stem of a tree.

tryma, *pl.* **trymata.**   A two- to four-loculed nut which is surrounded at maturity, by a dehiscent involucre, e.g. *Carya*.

tube.   The more or less cylindrical portion of a gamopetalous corolla or a gamosepalous calyx.

tube cell.   **1.** One of the cells resulting from the germination of the microspore in gymnosperms and flowering plants; the cell that develops into the pollen tube. **2.** An elongated cell with lignified walls found in the inner epidermis of the pericarp of a caryopsis (grass, Poaceae).

tube nucleus.   The nucleus of the tube cell which may or may not direct the development of a pollen tube. Normally located near the tip of the growing tube. (*See* fig. P-17).

tuber.   A thickened, compressed, fleshy, usually underground stem; functions include food and water storage, and propagation. (*See* fig. T-8).

tubercle.   A small, tuberlike swelling, nodule or projection.

tuberculed, tuberculate.   Having tubercles.

tuberiferous.   Bearing tubers.

tuberous.   Having tubers or resembling a tuber.

tubular.   Having the shape of a hollow cylinder with more or less parallel sides.

tufted.   In clumps; clustered, cespitose.

tumid.   Swollen or inflated.

tumor.   A mass of tissue which grows independently of surrounding tissues and may invade those tissues.

tundra.   A low growing vegetation type occurring in cold regions of the world where the soil is frozen much of the year; the vegetation physiognomy is mostly xeromorphic, owing to the extremes of wind exposure and low humidity.

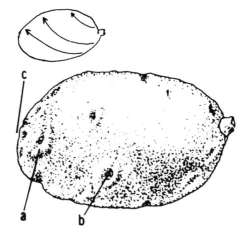

Fig. T-8. Tuber: a. bud scale scar, b. lateral bud, ''eye,'' c. apical end.

tunic. A loose membranous outer skin which does not develop from the epidermis.

tunica-corpus concept. A concept of organization of the shoot apical meristem of most angiosperms and a few gymnosperms. It consists of one or more peripheral layers of cells (the tunica layers) and the interior (the corpus). The tunica layers undergo surface growth (by anticlinal divisions), and the corpus undergoes volume growth (by divisions in all planes).

tunicate, tunicated. Having concentric layers or coats; as the leaves of certain bulbs.

turbinate. Shaped like a top; inversely conical.

turgid. Swollen or inflated; said of a cell that is firm due to uptake of water.

turgor. The state of being turgid.

turgor movement. A movement that results from changes in turgor pressure.

turgor pressure. The pressure exerted against a cell wall as a result of the movement of water into the cell.

turion. **1.** A young shoot or sucker which is produced from an underground stem. **2.** A perennating winter bud which is separated from the parent and gives rise to a new plant in the spring, as in *Potamogeton*.

tussock. A tuft of grass or grasslike plants.

twig. A shoot or branch of a tree produced during the current growing season.

twining. Twisted around an axis.

two-lipped. *See* bilabiate.

two-trace unilacunar condition. A stem node which has two leaf traces associated with one leaf which has one leaf gap.

tylose, *pl.* **tyloses.** A bubblelike outgrowth from a ray or axial parenchyma cell, which extends into the lumen of a vessel through a pit cavity.

tylosoid. Cellular outgrowths which resemble tyloses; e.g. outgrowths of parenchyma cells into sieve elements, and of epithelial cells of some gymnosperms which close (clog) resin or gum ducts.

type, type specimen. **1.** *See* holotype. **2.** Other types include isotype, syntype, lectotype, neotype and paratype.

typology.   An old view which disregarded population variation and considered members of a population as copies of the holotype.

typonym.   A synonym; a name based on the same type, not on a diagnosis or description.

U

ubiquitous. Present or occurring everywhere.

ultramicrotome. A microtome designed to cut extremely thin tissue sections, usually for examination with an electron microscope.

ultraviolet floral pattern. A pattern of ultraviolet reflection and absorption on a flower or inflorescence. (*See* fig. U-1).

ultraviolet radiation (UV). A portion of the electromagnetic spectrum which lies beyond the violet at wavelengths between 100 and 400 nm. An important mutogenic and sterilization agent.

umbel. A determinate or indeterminate flat-topped or convex inflorescence in which the pedicels all arise from the apex of the peduncle. (*See* fig. P-1).

umbellate. Borne in umbels; in the form of an umbel.

Fig. U-1. Ultraviolet floral patterns: A. U.V. reflective area (white), B. U.V. absorptive area (dark).

umbellet.   A secondary umbel of a compound umbel. *Syn.* umbellule.

umbelliferous.   Bearing umbels.

umbelliform.   Resembling an umbel, umbel-shaped.

umbilicate.   Having a depression in the center.

umbilicus.   The hilum of a seed.

umbo, *pl.* **umbones.**   A blunt or rounded projection arising from a surface; as on a pine cone scale. *Syn.* boss.

umbonate.   Bearing an umbo.

umbonulate.   Bearing a very small umbo.

umbraculate.   Having the shape of an umbrella.

unarmed.   Lacking any projections, such as prickles, spines or awns; blunt or muticous.

uncinate.   Hooked at the tip. (*See* fig. U-2).

unctuous.   Oily or greasy to the touch.

underleaf.   A small leaf on the underside of the stem of liverworts.

understory.   The trees in a forest found beneath the level of the main canopy.

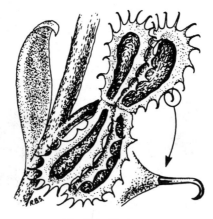

Fig. U-2. Uncinate.

undulate.   A margin wavy (up and down) in the vertical plane; less pronounced than sinuate.

unequal crossing-over.   Improper pairing of adjacent, duplicated sequences which are in tandem order. Crossing-over in these regions results in one chromatid with one copy of duplicated sequence and one chromatid with three copies of the sequence.

unguiculate.   Contracted at the base into a claw, forming a stalk.

uni-.   A prefix meaning one or single.

uniaxial.   A type of growth in which the main axis consists of a single filament of usually large cells; as in certain red algae (Rhodophyta). Compare multiaxial.

unicarpellate, unicarpellous.   Having a single carpel in the fruit.

unicellular.   Composed of a single cell.

unifacial leaf.   A leaf which has a similar anatomical structure on both sides. Compare bifacial leaf.

unifoliolate.   A compound leaf which is reduced to a single, usually terminal, leaflet. The petiolule is distinct from a petiole.

unilacunar node.   A node with one leaf gap per leaf.

unilateral.   One-sided.

unilocular.   Having a single locule or chamber. (*See* fig. P-12).

unilocular zoosporangia.   Unicellular zoosporangia on diploid plants in which meiosis occurs producing haploid zoospores; e.g. as in *Ectocarpus* (Phaeophyta). Compare plurilocular. (*See* fig. P-12).

uniparental reproduction.   Organisms which habitually reproduce asexually, forming clones; frequently excluded from the concept of biological species on the basis that they do not form breeding populations.

uniseriate.   Having a single, horizontal row or series of cells.

uniseriate ray.   In secondary vascular tissues, a ray one cell wide.

unisexual.   **1.** Applied to flowers that have either stamens or pistil(s) but not both in the same flower. **2.** A gametophyte that produces only antheridia or archegonia but not both.

unit cup fruit.   A fruit derived from a flower having an inferior ovary with united carpels. Fruit wall consists of ovary wall (pericarp) and accessory parts.

unitegmic ovule.   An ovule with a single integument.

unit free fruit.   A fruit derived from a flower having a superior ovary with united carpels. Fruit wall consists of pericarp only.

unit membrane.   A hypothetical concept that describes the basic membrane structure of most membranes (e.g. plasma, tonoplast, and most cellular organelles) as consisting of three layers forming a unit; as seen with the electron microscope, two dark lines are separated by a clear space.

unitunicate.   An ascus in which both the inner and outer wall are more or less rigid and do not separate or expand during spore ejection.

univalent.   An unpaired chromosome during meiosis.

universal veil.   A thin, veillike membrane which covers the developing basidiocarp in the Agaricales (Basidiomycetes). The remnants are called the volva.

upright ray cell.   In secondary vascular tissues, a ray cell having its longest dimension oriented vertically to the axis.

upwelling.   An oceanic process in which winds move surface water away from coastal slopes, thereby bringing to the surface, cold water rich in nutrients.

urceolate.   Urn-shaped, constricted just below the mouth.

urediniospore.   *See* uredospore.

uredinium, *pl.* **uredinia.**   The structure that produces uredospores in the rust fungi (Basidiomycetes). *Syn.* uredium, *pl.* uredia.

urediospore.   *See* uredospore.

uredium.   *See* uredinium.

uredospore.   A reddish, binucleate spore produced by rust fungi (Basidiomycetes). It may reinfect the same host species on which it is produced. *Syn.* urediospore or urediniospore.

urn.   The base of a pyxis.

urticating hairs.   The stinging hairs of nettles (Urticaceae) and other plants.

utricle.   A small bladder; a one-seeded, dry fruit, often dehiscent by a lid.

utricular.   Inflated or bladderlike.

UV.   *See* ultraviolet radiation.

V

vacuolar membrane. *See* tonoplast.

vacuole. A membrane-bound region within the cytoplasm of certain plant cells; they are filled with a watery liquid, the cell sap, rather than with protoplasm. They function in maintenance of water balance within the cell, accumulation of various ions and molecules, storage of pigments, etc.

vacuome. A collective term for all vacuoles in a cell, tissue or plant.

vaginate. Sheathed.

valence. A number which represents the capacity of an element or radical to combine with another to form molecules. The number of electrons gained, lost or shared by an atom in a compound or the number of hydrogen or chlorine atoms which one radical or atom of an element will combine with or replace.

vallecular. Pertaining to the grooves between the ridges; as in fruits of the umbel family (Apiaceae).

vallecular canals. Air-containing canals which alternate with the vascular bundles in the stems of horsetails (*Equisetum*).

valvate. **1.** Opening by valves. **2.** Edges of structures coming together so that the margins touch but do not overlap. Compare imbricate.

valve. **1.** In diatoms and some dinoflagellates one of the two halves of a cell wall or frustule. **2.** In anthers which open by pores, the flap of anther tissue which covers the pore. **3.** In fruits, one of the units or parts into which the fruit wall separates at maturity, e.g. as in a capsule or legume.

vapor. The gaseous state of a substance that is normally a liquid or a solid.

variance. The mean of the squared deviations for all values, both positive and negative, from the population mean. The square of the standard deviation.

variant. Any definable individual or group which is different.

variate. A single observation or measurement which is under investigation in a statistical study.

variation. The divergence among individual progeny of a particular group or species.

Such differences are due to the genetic constitution of the individual responding to environmental conditions.

varicose.   Abnormally and irregularly enlarged or swollen.

variegated.   Descriptive of leaves which lack chlorophyll in certain sections, thus appearing yellowish or white; also in reference to flowers, seeds, etc. in which pigmentation is not uniform in intensity. (*See* fig. V-1).

variety.   A subdivision of a species which differs as a group in some minor definable characteristic(s) from the rest of the species.

vascular.   Refers to any plant tissue or area which consists of, or gives rise to conducting tissue, e.g. xylem, phloem or vascular cambium.

vascular bundle.   A strand of vascular tissue composed of xylem, phloem and procambium (if still present). (*See* figs. C-8, H-6, R-5).

vascular bundle sheath.   *See* bundle sheath.

vascular cambium.   A lateral meristematic region that gives rise to secondary phloem and xylem in the stem and root. Lies between phloem and xylem. (*See* fig. C-12).

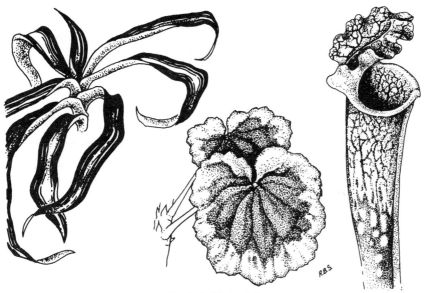

Fig. V-1. Variegated.

vascular cylinder. The vascular tissue and associated ground tissues of a stem or root. *Syn.* stele.

vascular plant. Any plant which contains vascular tissues (xylem and phloem); e.g. a member of the Tracheophyta, which includes the seedless vascular plants and the seed plants (gymnosperms and angiosperms).

vascular rays. Ribbonlike sheets of parenchyma, produced by the vascular cambium. They extend radially from the secondary xylem across the cambium and into the secondary phloem.

vascular tissue. The conducting tissues of a plant composed of primary and/or secondary xylem and phloem. Includes procambium and/or vascular cambium.

vascular tissue system. Refers to all the vascular tissues in a plant or plant organ.

vascular trace. *See* trace.

vascular tracheids. Cells which resemble vessel members in form and arrangement but lack perforations; regarded as incompletely developed vessel members.

vasculum. A metal container used to temporarily store plants when collecting botanical specimens in the field. Largely replaced by the plastic bag.

vasicentric. Concentrated around the vessels.

vasicentric paratracheal parenchyma. Axial parenchyma associated with vessels in secondary xylem which form a complete sheath around a vessel. *See* paratracheal parenchyma.

vasiform. Having the shape of an elongated funnel.

vector. A carrier; an organism, usually an insect, that carries pollen or disease-causing organisms (pathogens) from one plant to another.

vegetable. Of or pertaining to a plant or plant part. Frequently used in reference to an edible portion of a plant which is not derived as a product of sexual reproduction, i.e. not a fruit.

vegetation. The sum total of the plants growing in an area or region.

vegetative. Pertaining to plant organs or parts which have nonreproductive functions, e.g. leaves, roots, stems, etc.; somatic tissue.

vegetative apomixes. *See* vegetative propagation.

vegetative cell.   The tube cell and its included haploid nucleus in a pollen grain.

vegetative growth.   Growth of vegetative organs, e.g. leaves, roots, stems, etc. and other somatic tissues.

vegetative propagation, vegetative reproduction.   Asexual propagation; propagation by vegetative tissues or organs; e.g. leaves, stems, tubers, rhizomes, stolons, etc. Unless a mutation occurs, all progeny from such reproduction are genetically identical to the parents. *Syn.* vegetative apomixis.

vegetative reproduction.   *See* vegetative propagation.

vein.   A strand of vascular tissue (a vascular bundle), which is part of the network of supporting and conducting tissue of an expanded organ, as a leaf, petal, etc.

veinlet.   A small vein.

velamen.   A multiple epidermis on the aerial roots of some orchids and aroids, which functions in water absorption. Also found on some terrestrial roots.

velum.   1. A veil or hyphal membrane which extends from the stalk to the pileus in some fungi; e.g. death angel (*Amanita*). 2. A membranous protective flap or indusium covering the sporangium, as in quillwort (*Isoetes*). 3. A membraneous structure involved in the closure of the trap in bladder pod (*Utricularia*).

velutinous.   Covered with fine, dense, straight, long and soft trichomes; velvety.

venation.   The arrangement of the vascular bundles or veins in a leaf blade.

veneer.   A thin sheet of wood, used to make plywood or as a furniture veneer; normally made by rotating a log or bolt against a knife in a lathe. (*See* fig. V-2).

venter.   The enlarged basal portion of an archegonium containing the egg. (*See* fig. A-18).

ventral.   Pertaining to the surface nearest the axis; the upper surface of a leaf; the inner surface of an organ. *Syn.* adaxial. *Opp.* dorsal.

ventral canal cell.   A cell produced by mitotic division of the central cell, and which is situated over its sister cell (the egg), in the venter of archegonia; as in lycopods and ferns.

ventricose.   Enlarged or swollen unequally, that is, more on one side than the other; inflated on one side near the middle. *Syn.* gibbose.

venulose.   Having many veinlets.

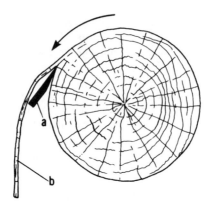

Fig. V-2. Veneer: a. cutting blade, b. veneer sheet.

vermicular, vermiform.   Worm-shaped or wormlike.

vermiculate.   Marked with impressions, as if worm-eaten.

vernal.   Pertaining to or appearing in the spring.

vernalization.   The natural or artificial induction of early flowering by exposure of seeds to low temperatures; also, the requirement for breaking bud dormancy of certain temperate woody perennials.

vernal pool.   A temporary pool usually formed during the rainy season; frequently supports a unique flora and fauna.

vernation.   The manner in which individual leaves are arranged in a bud before it expands; e.g. circinate vernation. *Syn.* ptyxis. Compare aestivation.

vernicose.   Shiny as though varnished.

verruca, *pl.* **verrucae.**   Wartlike processes on pollen grains.

verrucate.   Pollen with verruca always broader than high, and exceeding one micrometer in length; warty.

verrucose.   Warty; covered with wartlike outgrowths.

versatile.   An anther attached near its middle to the apex of the filament, and capable of swinging more or less freely.

versicolor.   Variegated; variously colored.

vertical parenchyma. *See* axial parenchyma.

verticil. A whorl; a whorled arrangement of similar parts, e.g. flowers.

verticillaster. A false whorl composed of a pair of nearly sessile, determinate, cymose inflorescences (dichasia) found at the nodes, as in the mints (Lamiaceae).

verticillate. Borne in verticils.

vesicle. A small sac, or cavity; a spherical body.

vesicular. Pertaining to, having or composed of vesicles.

vesiculate. Becoming vesicular; bladderlike.

vespertine. Appearing, blossoming, opening or functioning in the evening.

vessel. The water and mineral conducting structure of the xylem tissue of nearly all angiosperms; it is a tubelike structure of indeterminate length, composed of specialized nonliving cells (vessel members) placed end to end; end walls of the vessel members are open (perforated) forming simple or multiperforate perforation plates; usually numerous bordered pit-pairs occur between contiguous vessel members. (*See* figs. C-8, V-3).

vessel cell, vessel element. *See* vessel member.

vessel member. One of the cells that make up a vessel. *Syn.* vessel cell, vessel element.

vestibulum. A cavity inside a porus of a pollen grain, which is caused by a separation of two layers of the exine.

vestige, vestigial. Rudimentary; pertaining to an undeveloped or degenerate structure which is no longer functionally useful.

vestiture. Any covering on a surface, e.g. plant trichomes. *Syn.* vesture.

vesture. *See* vestiture.

vestured pit. A bordered pit which has projections facing into the cavity from the overhanging secondary wall.

vexillate. Having a standard or banner petal, as in a papilionaceous flower of the legumes (Fabaceae).

vexillum, *pl.* **vexilla.** *See* banner.

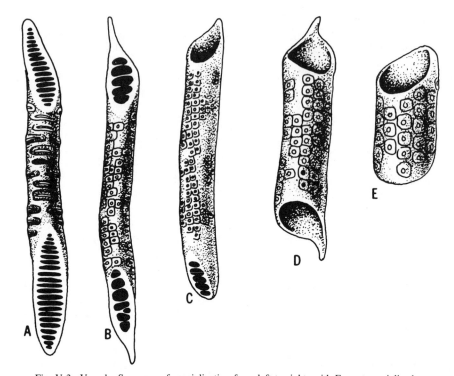

Fig. V-3. Vessels: Sequence of specialization from left to right, with E most specialized.

viable.   Able to survive and develop; as with viable seeds that remain alive and are able to germinate.

vibratile pollination.   Pollination which requires the direct mechanical vibration of the anther for the emission of the pollen. Several species of bees are capable of using their indirect flight muscles to vibrate the anthers. *Syn.* buzz pollination.

vibrio.   Any short, flagellate, gram-negative, cresent or S-shaped bacterial cell; e.g. species in the genus *Vibrio*.

vicariads.   Closely related taxa which replace each other along a geographical gradient.

vicinism.   Cross-fertilization predominately between neighboring individuals.

villose, villous.   Covered with long, soft, fine trichomes.

villosulous.   Minutely villous.

villus, *pl.* **villi.**   A long, soft, fine trichome.

vine.   An elongate, weak, often climbing stem.

virgate.   Wandlike, slender, straight and erect.

viridescent.   Becoming green.

viroid.   The smallest known agent of infectious disease. Conventional viruses are made up of nucleic acid with a protein coat, whereas viroids have only the nucleic acid.

virulence.   The capacity or ability of an organism to produce disease.

virus.   A submicroscopic, noncellular entity, composed of a nucleic acid core (DNA or RNA) encased in a protein shell. They reproduce only within host cells.

viscid, viscous.   Sticky; glutinous.

viscidulous.   Slightly viscid.

viscin.   In certain orchids, interconnecting threads between small groups of pollen grains (massulae).

vital stain.   Any dye which stains living material.

vitamin.   A complex organic compound which is necessary, in minute amounts, for normal growth; plants synthesize all of the vitamins necessary for their metabolic processes.

vitta.   An oil tube; common in the pericarp of most members of the umbel family (Apiaceae).

viviparous, vivipary.   Germination of seed while still attached to the parent plant, e.g. mangrove. (*See* fig. V-4).

volatile.   Evaporating or vaporizing readily; as the volatile oils of plants.

voluble.   Twining.

volute.   Rolled up, spiraled.

volva.   A cup-shaped structure, the remnant of universal veil, at the base of the stalk (stipe) of some mushrooms (Agaricales). (*See* fig. B-2).

volvocine line.   An evolutionary series in the green algae (Chlorophyta) consisting of colonial forms; e.g. *Volvox, Gonium.*

voucher.   A specimen preserved for future reference.

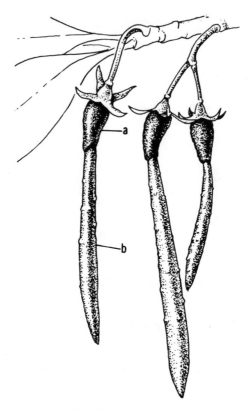

Fig. V-4. Viviparous: a. seed, b. radicle.

W

wall. *See* cell wall, primary wall, secondary cell wall.

wall pressure. The pressure of the cell wall exerted against the turgid cell protoplast. The opposing force is turgor pressure.

water balance. The difference between the amount of water absorbed by a plant and the amount lost through transpiration.

water bloom. *See* bloom.

water deficit. A negative water balance in a plant; results from a greater loss of water through transpiration than is gained through absorption.

water potential. The chemical potential of water (free energy per mole); the algebraic sum of osmotic potential and the pressure potential (wall pressure). Represented by the Greek letter psi, Ψ.

watershed. A geographical region which drains into a particular body of water.

water table. A subsurface soil level which is completely saturated with water.

water vesicle. An enlarged, highly vacuolated epidermal cell. A type of trichome in which water is stored.

waxes. A class of fats derived from fatty acids and alcohols other than glycerol; most waxes are obtained from seeds and also from leaves and fruits, where they help reduce water loss.

webbing. A hypothetical step in the evolutionary development of the leaf (megaphyll), in which the apical meristems of separate branches coalesced to produce a webbed structure of parenchyma tissue. *See* Telome theory.

weed. A plant, usually herbaceous, which is growing in an area where it is neither desired nor appreciated.

wetting agents. Chemicals added to water which reduce or break the surface tension.

whiplash flagellum. A flagellum without mastigonemes. Compare tinsel flagellum. (*See* fig. W-1).

whole arm fusion.   A reciprocal translocation involving two rod-shaped, telocentric chromosomes; an entire chromosome arm is transferred resulting in a reduction in chromosome number.

whorl.   1. A group of three of more parts at a node; e.g. leaves or branches. **2.** A circle of floral organs; e.g. stamens, petals, carpels.

wild type.   In genetics, the most frequently observed phenotype, which is characteristic of a species in a natural environment.

wilting.   The loss of turgor pressure in a plant, so that it becomes limp or droopy. *See* permanent wilting percentage.

wind-pollinated.   A plant which is pollinated by means of wind transported pollen.

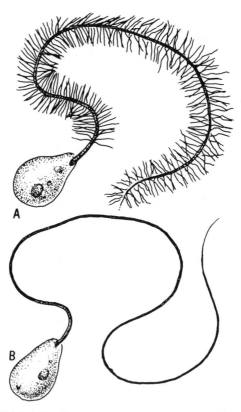

Fig. W-1. Whiplash flagellum: A. tinsel flagellum, B. whiplash flagellum.

wing.   1. A thin, flat, membraneous or leathery appendage on the surface of an organ, e.g. a seed or fruit. **2.** One of the two lateral petals of a papilionaceous corolla.

winter annual.   An annual which usually germinates in the late fall, grows vegetatively during the winter, and sets flowers and fruits in early spring.

winter hardiness.   The ability of a plant to tolerate severe winter conditions.

wood.   Technically, the secondary xylem of gymnosperms and dicotyledons; but also applies to any other xylem. *See* secondary xylem.

wood fiber.   *See* xylem fiber.

woody.   Refers to plants or plant parts which have hard and lignified secondary tissue.

woolly.   Covered with long matted hairs. *Syn.* tomentose.

wound periderm.   Periderm which is formed in response to wounding or other injury.

wound tissue.   *See* callus.

X

x. *See* basic number.

xanthophyll. Yellow or brownish pigments of the carotenoid group, which are closely related to carotene. They occur in leaf chloroplasts (e.g. lutein, zeaxanthin), in brown algae (e.g. fucoxanthin) and diatoms (e.g. diatoxanthin).

x-axis. The horizontal axis of a graph.

x chromosome. The sex chromosome found in a double dose in the homogametic sex and in a single dose in the heterogametic sex.

xenia. The direct effect of pollen on the embryo and endosperm tissue. Compare metaxenia.

xenogomy. Cross-pollination involving flowers on different plants.

xeric. Dry, without moisture. Adapted to dry or desert conditions. Compare mesic.

xero-. A prefix meaning dry.

xeromorphic. Refers to special or typical morphological adaptations of xerophytes; having a form that is structurally adapted for growth under xeric conditions.

xerophyte. A plant that is adapted to dry or arid habitats.

x number. *See* basic number.

xylary fibers. Fibers found within the xylem tissue. Compare extraxylary fibers.

xylem. The principle water- and mineral-conducting tissue in vascular plants; a complex tissue composed of nonliving, lignified tracheids, vessels, and fibers, and their associated parenchyma cells. Xylem may also provide mechanical support, especially in plants with secondary xylem (wood). (*See* figs. A-23, C-4, M-4).

xylem elements. The elongated to drum-shaped cells which, together with tracheids, are the two fundamental types of tracheary elements in the xylem.

xylem fiber. A fiber of the xylem tissue that intergrades morphologically with tracheids and with parenchyma cells. Two types are recognized in the secondary xylem: fiber-tracheids and libriform fibers. *Syn.* wood fiber.

xylem member.   *See* xylem elements.

xylem ray.   The portion of a vascular ray which is located in the secondary xylem; it is composed of ray parenchyma cells produced by the vascular cambium.

xylotomy.   The anatomy of xylem.

Y

y-axis. The vertical axis of a graph.

y chromosome. A sex chromosome found only in the heterogametic sex.

yeast. A unicellular, ascomycete fungus capable of anerobic respiration; many are important to man because of their ability to ferment carbohydrates, e.g. as in the production of alcohol and rising of bread dough.

yellows. A symptom of disease in plants in which a normally green plant part turns yellow or is chlorotic.

Z

zeatin. A plant hormone; one of the most active of the known cytokinins, chemically related to kinetin.

zonal soils. A profiled soil developed under specific climatic and vegetational conditions.

zonoaperturate. A pollen grain with apertures located in an equatorial zone; e.g. zonocolpate, zonocolporate, zonoporate, zonopororate.

zoochore. A plant which is disseminated by animals.

zooidogamy. Descriptive of plants which are fertilized by motile flagellated sperms; e.g. lower vascular plants. Compare siphonogamy.

zooplankton. Animal plankton.

zoosporangium. A sporangium in which flagellated or ciliated spores are produced.

zoospore. A motile, asexually produced spore, found among algae and fungi.

zygomorphic. Bilaterally symmetrical; especially in reference to a flower or corolla. *Opp.* actinomorphic. *Syn.* irregular flower.

zygomorphy. The condition of being bilaterally symmetrical.

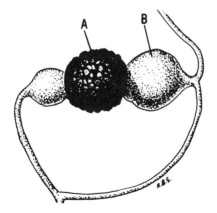

Fig. Z-1. Zygospore: A. mature zygospore, B. suspensor.

zygonema.   *See* zygotene.

zygophore.   A specialized, hyphal, club-shaped branch of certain Zygomycetes; zygophores of opposite mating types fuse and eventually form zygospores.

zygosporangium.   A sporangium which contains a zygospore.

zygospore.   A thick-walled resistant spore which develops from a zygote, following fusion of isogametes of gametangia; a resting spore. (*See* fig. Z-1).

zygote.   The single diploid (2n) cell formed by the fusion of male and female gametes (i.e. egg and sperm). Mitotic divisions of the zygote produce the embryo. (*See* fig. A-9).

zygotene.   The second stage in prophase I of meiosis; the homologous thread-like chromosomes become intimately associated throughout their lengths, and form a four-stranded structure (termed bivalent or tetrad); this pairing process is termed synapsis. *Syn.* zygonema.

Bibliography

Axelopoulos, Constantine John. 1962. *Introductory Mycology*. 2d ed. New York: John Wiley & Sons.

Allard, R. W. 1960. *Principles of Plant Breeding*. New York: Wiley.

Ambrose, E. J., and Easty, Dorothy M. 1970. *Cell Biology*. Reading, Mass.: Addison-Westley Publishing Co.

Andrews, Henry N., Jr., 1961. *Studies in Paleobotany*. New York: John Wiley & Sons, Inc.

Avers, Charlotte J. 1976. *Cell Biology*. New York: D. Van Nostrand Co.

Benson, Lyman. 1959. *Plant Classification*. Lexington, Mass.: D. C. Heath & Co.

Bierhorst, David W. 1971. *Morphology of Vascular Plants*. New York: The Macmillan Co.

Bold, Harold C. 1967. *Morphology of Plants*. 2d ed. New York: Harper and Row, Publishers.

Bold, Harold C. 1970. *The Plant Kingdom*. 3d ed. Foundations of Modern Biology Series. Englewood Cliffs, New Jersey: Prentice-Hall, Inc.

Briggs, D., and Walters, S. M. 1972. *Plant Variation and Evolution*. New York: McGraw-Hill Book Co.

Brock, Thomas D. 1970. *Biology of Microorganisms*. Englewood Cliffs, New Jersey: Prentice-Hall, Inc.

Carlquist, Sherwin. 1961. *Comparative Plant Anatomy*. New York: Holt, Rinehart & Winston.

Carlquist, Sherwin. 1975. *Ecological Strategies of Xylem Evolution*. Berkeley, Calif.: University of California Press.

Chamberlain, Charles Joseph. 1966. *Gymnosperms, Structure and Evolution*. New York: Dover Publications, Inc.

Chapman, Stephen R., and Carter, Lark P. 1976. *Crop Production Principles and Practices*. San Francisco: W. H. Freeman & Co.

Conard, H. S. 1956. *How to Know the Mosses and Liverworts*. Dubuque, Iowa: Wm. C. Brown Co., Publishers.

Cronquist, Arthur. 1968. *The Evolution and Classification of Flowering Plants*. Boston: Houghton Mifflin Co.

Cronquist, Arthur. 1971. *Introductory Botany*. 2d ed. New York: Harper and Row, Publishers.

Curtis, Helena. 1977. *Invitation to Biology*. 2d ed. New York: Worth Publishers, Inc.

Davis, P. H., and Heywood, V. H. 1973. *Principles of Angiosperm Taxonomy*. New York: Krieger Publishing Co.

Dean, H. L. 1978. *Laboratory Exercises. Biology of Plants*. 4th ed. Dubuque, Iowa: Wm. C. Brown Co., Publishers.

Dyson, Robert D. 1974. *Cell Biology, A Molecular Approach*. Boston: Allyn and Bacon, Inc.

Easu, Katherine. 1953. *Plant Anatomy*. New York: John Wiley and Sons, Inc.

Easu, Katherine. 1965. *Plant Anatomy*. 2d ed. New York: John Wiley and Sons, Inc.

Easu, Katherine. 1977. *Anatomy of Seed Plants*. New York: John Wiley and Sons, Inc.

Etherington, J. R. 1975. *Environment and Plant Ecology*. New York: John Wiley & Sons, Inc.

Faegri, K., and van der Pijl, L. 1971. *The Principles of Pollination Ecology*. 2d ed. New York: Pergamon Press.

Fahn, A. 1974. *Plant Anatomy*. 2d ed. Oxford: Pergamon Press.

Featherly, H. I. 1954. *Taxonomic Terminology of the Higher Plants*. New York: Hafner Publishing Co.

Fernald, Merritt Lyndon. 1950. (Corrected Printing, 1970 by Rollins), *R. C. Gray's Manual of Botany.* 8th (Centennial) ed. - Illus. New York: D. Van Nostrand Co.

Foster, Adriance S. 1949. *Practical Plant Anatomy.* New York: D. Van Nostrand Co., Inc.

Foster, Adriance S., and Gifford, Ernest M. Jr., 1974. *Comparative Morphology of Vascular Plants.* San Francisco: W. H. Freeman and Co.

Fuller, Harry J., and Ritchie, Donald D. 1967. *General Botany.* 5th ed. New York: Barnes & Noble Books.

Fuller, Harry J., Carothers, Zane B., Payne, Willard W., and Blaback, Margaret K., 1972. *The Plant World.* New York: Holt, Rinehart and Winston, Inc.

Goldsby, Richard A. 1976. *Biology.* New York: Harper and Row, Publishers.

Grant, Verne. 1971. *Plant Speciation.* New York: Columbia University Press.

Grant, Verne. 1975. *Genetics of Flowering Plants.* New York: Columbia University Press.

Grant, Verne. 1977. *Organismic Evolution.* San Francisco: W. H. Freeman and Co.

Greulach, Victor A. 1973. *Plant Function and Structure.* New York: The Macmillan Co.

Greulach, Victor A., and Adams, J. Edison, 1967. *Plants: An Introduction to Modern Botany.* New York: John Wiley and Sons, Inc.

Hardin, Garrett, and Bajema, Carl. 1978. *Biology and its Implications,* San Francisco: W. H. Freeman and Co.

Hartmann, Hudson, T., and Kester, Dale E. 1975. *Plant Propagation, Principles and Practices.* Englewood Cliffs, New Jersey: Prentice-Hall, Inc.

Herskowitz, Irwin J. 1965. *Genetics.* 2d ed. Boston: Little, Brown and Co.

Hitchcock, C. Leo, and Cronquist, Arthur. 1973. *Flora of the Pacific Northwest, An Illustrated Manual.* Seattle: University of Washington Press.

Janick, Jules. 1972. *Horticultural Science.* San Francisco: W. H. Freeman and Co.

Janick, Jules, Schery, Robert W., Woods, Frank W., and Rutan Vernon W. 1969. *Plant Science. An Introduction to World Crops.* San Francisco: W. H. Freeman and Co.

Jensen, William A., and Salisbury, Frank B. 1972. *Botany: An Ecological Approach.* Belmont, California: Wadsworth Publishing Co., Inc.

Jones, Kenneth C., and Gaudin, Anthony J. 1977. *Introductory Biology.* New York: John Wiley & Sons, Inc.

Kenneth, John H. 1963. *Henderson's Dictionary of Biological Terms.* 8th ed. New York: D. Van Nostrand Co., Inc.

Lawrence, George H. M. 1951. *Taxonomy of Vascular Plants.* New York: The Macmillan Co.

Leopold, A. Carl, and Kriedmann, Paul E. 1975. *Plant Growth and Development.* 2d ed. New York: McGraw-Hill Book Co.

Lincoff, Gary, and Mitchell, D. H. 1977. *Toxic and Hallucinogenic Mushroom Poisoning. A Handbook for Physicians and Mushroom Hunters.* New York: Van Nostrand Reinhold Co.

Linskens, H. F., ed. 1974. *Fertilization in Higher Plants.* North-Holland Publishing Co., Amsterdam, Oxford. New York: American Elsevier Publishing Co., Inc.

Lott, John N. A. 1976. *A Scanning Electron Microscope Study of Green Plants.* St. Louis, Missouri: The C. V. Mosby Co.

MacArthur, Robert H., and Wilson, Edward O. 1967. *The Theory of Island Biogeography.* Princeton, New Jersey: Princeton University Press.

Mahlberg, Paul. 1972. *Laboratory Program in Plant Anatomy.* Dubuque, Iowa: Wm. C. Brown Co., Publishers.

Merrel, David J. 1975. *An Introduction to Genetics.* New York: W. W. Norton and Co., Inc.

Meyer, Bernard S., Anderson, Donald B., and Bohning, Richard H. 1973. *Introduction to Plant Physiology.* New York: D. Van Nostrand Co., Inc.

Miller, G. Tyler. 1975. *Living in the Environment. Concepts, Problems and Alternatives.* Belmont, California: Wadsworth Publishing Co.

Moore, D. M. 1976. *Plant Cytogenetics*. A Halsted Press Book, New York: John Wiley & Sons, Inc.

Mueller-Dombois, Dieter, and Ellenberg, Heinz. 1974. *Aims and Methods of Vegetation Ecology*. New York: John Wiley & Sons, Inc.

Munz, Philip A. 1974. *A Flora of Southern California*. Berkeley, California: Univ. of California Press.

Noggle, G. Ray, and Fritz, George J. 1976. *Introductory Plant Physiology*. Englewood Cliffs, New Jersey: Prentice-Hall, Inc.

O'Brien, T. P., and McCully, Margaret E. 1969. *Plant Structure and Development. A Pictorial and Physiological Approach*. London: The Macmillan Co. Collier-Macmillan, Limited.

Odum, Eugene P. 1971. *Fundamentals of Ecology*. 3d ed. Philadelphia: W. B. Saunders Co.

Oosting, Henry J. 1948. *The Study of Plant Communities. An Introduction to Plant Ecology*. 2d ed. San Francisco: W. H. Freeman and Co.

Payne, W. W. 1978. A Glossary of Plant Hair Terminology. *Brittonia* 30:239-255.

Porter, C. L. 1959. *Taxonomy of Flowering Plants*. San Francisco: W. H. Freeman and Co.

Proctor, Michael, and Yeo, Peter. 1971. *The Pollination of Flowers*. New York: Taplinger Publishing Co.

Radford, Albert E., Dickison, William C., Massey, Jimmy R., and Bell, C. Ritchie. 1974. *Vascular Plant Systematics*. New York: Harper and Row, Publishers.

Raven, Peter H., Evert, Ray F., and Curtis, Helena. 1976. *Biology of Plants*. New York: Worth Publishers, Inc.

Ray, Peter Martin. 1972. *The Living Plant*. New York: Holt, Rinehart and Winston, Inc.

Ricklefs, Robert E. 1973. *Ecology*. Newton, Massachusetts: Chiron Press, Inc.

Ruchforth, Samuel R. 1976. *The Plant Kingdom. Evolution and Form*. Englewood Cliffs, New Jersey: Prentice-Hall, Inc.

Salisbury, Frank B., and Ross, Cleon. 1969. *Plant Physiology*. Belmont, California: Wadsworth Publishing Co., Inc.

Scagel, Robert F., Bandoni, Robert J., Rouse, Glenn E., Schofield, W. B., Stein, Janet R., and Taylor, T. M. C. 1965. *An Evolutionary Survey of the Plant Kingdom*. Belmont, California: Wadsworth Publishing Co., Inc.

Scagel, Robert F., Bandoni, Robert J., Rouse, Glenn E., Schofield, W. B., Stein, Janet R., and Taylor, T. M. C. 1965. *Plant Diversity: An Evolutionary Approach*. Belmont, California: Wadsworth Publishing Co.

Schery, Robert W. 1972. *Plants for Man*. 2d ed. Englewood Cliffs, New Jersey: Prentice-Hall, Inc.

Sienko, Michell J., and Plane, Robert A. 1961. *Chemistry*. New York: McGraw-Hill Book Co., Inc.

Stanfield, William D. 1977. *The Science of Evolution*. Macmillan Publishing Co., Inc., New York. London: Collier Macmillan Publishers.

Stebbins, G. Ledyard. 1966. *Processes of Organic Evolution*. Englewood Cliffs, New Jersey: Prentice-Hall, Inc.

Stebbins, G. Ledyard. 1971. *Chromosomal Evolution in Higher Plants*. Reading, Massachusetts: Addison-Wesley Publishing Co.

Steen, Edwin B. 1971. *Dictionary of Biology,* New York: Barnes & Noble Books, A Division of Harper & Row, Publishers.

Swanson, Carl P. 1957. *Cytology and Cytogenetics*. Englewood Cliffs, New Jersey: Prentice-Hall, Inc.

Swartz, Delbert. 1971. *Collegiate Dictionary of Botany*. New York: The Ronald Press Co.

Tippo, Oswald, and Stern, William Louis. 1977. *Humanistic Botany*. New York: W. W. Norton and Co., Inc.

Van der Pijl, L., and Dodson, Calaway H. 1966. *Orchid Flowers. Their Pollination and Evolution*. Coral Gables, Florida. Published Jointly by the Fairchild Tropical Garden and the University of Miami Press.

Walters, Dirk R. 1975. *Vascular Plant Taxonomy. A Study Guide*. San Luis Obispo, California: Toyon Publishing.

Webster, John. 1970. *Introduction to Fungi*. Cambridge: Cambridge University Press.

Weier, T. Elliot, Stocking, C. Ralph, and Barbour, Michael G. 1970. *Botany. An Introduction to Plant Biology*. New York: John Wiley and Sons, Inc.

Westcott, Cynthia. 1971. *Plant Disease Handbook*. New York: Van Nostrand Reinhold Co.

Wilson, Carl L., Loomis, Walter E., and Steeves, Taylor A. 1971. *Botany*. 5th ed. New York: Holt, Rinehart and Winston.

White, Michael J. D. 1978. *Modes of Speciation*. San Francisco: W. H. Freeman and Co.

Zar, Jerrold H. 1974. *Biostatistical Analysis*. Englewood Cliffs, New Jersey: Prentice-Hall, Inc.